APPROACHABLE OPEN SOURCE

ian Muenzenmeyer

eword by Brad Frost

The views and opinions expressed in this book are solely those of the author and do not necessarily reflect the official policy or position of any employer, organization, open source project, or other group or individual.

Publisher: Brian Muenzenmeyer

Editors: Lisa Maria Marquis, Tina Lee

Technical Editor: Mat "Wilto" Marquis

Book cover by Brian Muenzenmeyer using css-doodle and codepen.io

Headings set in Urbanist by Corey Hu. Body content set in EB Garamond by Georg Mayr-Duffner

More info, updates, resources, and errata at approachableopensource.com

1.0.1 edition (semver), 2024

1 3 5 7 9 10 8 6 4 2

ISBN 979-8-9907798-0-8 (print) 979-8-9907798-1-5 (digital)

Library of Congress Control Number: 2024910658

Created in Chanhassen, Minnesota

Contents

To Megan, my caring, Luddite wife, who begged me not to ask her to read this.

If she ever opens a copy, she'll know she's the source of all the joy in my life.

Foreword

The World Wide Web is one of humanity's greatest achievements, and despite its bumpy ride the ideals and promise of the web are still very much alive. People connect, collaborate, build community, and work together to make the world a better place. Perhaps nothing better embodies the ideals of the web like open source software. Every day, millions of people from all over the world author, edit, delete, fix, and improve code across countless open source projects. We're collaborating at a scale the world has never seen before, which is absolutely wild when you stop and think about it.

I've never found the mechanics of open source software to be particularly sexy; git continues to give me grief despite years of me using it. But behind all of our commits, hashes, messages, issues, patches, releases, dependencies, and packages—are *people*. Open source software is *people*.

As you'll soon find out, Brian's enthusiasm and commitment to open source software is downright infectious. I got to witness Brian's tenacious open source spirit first hand through a design system tool called Pattern Lab I created in 2013 with Dave Olsen. A few months into the project, this guy with a complicated last name and GitHub handle made an unassuming entrance: Brian Muenzenmeyer simply commented on one of Pattern Lab's many GitHub issues to make a suggestion for getting the software to work on Windows machines. That's how these things go; sometimes important people make subtle entrances into your life that can only be perceived as profound with the power of hindsight. Brian quickly graduated from commenter to contributor to maintainer. Brian created a vision for Pattern Lab, helped us modernize the architecture, and instilled a real sense of open source maturity. Moreover, he injected so much enthusiasm, passion, love, and commitment into the project that inspired us to do our best.

But it wasn't all GitHub stars and accolades; Brian was maintaining and advancing Pattern Lab while raising four children—FOUR CHILDREN!—*and* working a full-time job. The double-edged sword that is passion caused him burn out; something I feel guilty about to this day. I wish I had this book on hand so I could have promoted a healthier culture around the project. But Brian's hardships have yielded incredible insights, which he graciously shares with us in this book. He acquired a metric ton of lemons and decided to make us all some fresh-squeezed lemonade; that just shows what kind of person he is.

I'm sorry to disappoint you if you were expecting a dry, mechanical book on the technical details of open source software. Don't get me wrong, Brian covers some real meat; this book is equal parts practical and profound. It is a much-needed handbook for navigating the waters of open source, and it's also a call to arms: for individuals to engage with open source, for communities to be open and inviting, for organizations to fund and support open source work. And Brian delivers it all in a way that's so incredibly human, poetic, inspirational, funny, and poignant.

This book will reshape how you think about open source, and I have no doubt Brian's words will inspire you to lean into the ideals of the web and contribute to something bigger than yourself. If you see something that's broken, you can fix it. If you have an idea, you can share it with the world. If you like what others are doing, you can collaborate with them. And maybe you'll get lucky like I did and those collaborators will become your friends. I can thank open source software for the gift that is my longstanding friendship with Brian Muenzenmeyer. And we can now thank open source software and Brian for the gift that is this incredible book you're about to read.

—Brad Frost

Chapter One

Principles and Promise

I'm a verbose communicator. My coworkers started creating flashcards of terms I introduce to them. Like *bifurcate*, *precipitous*, or *germane*. So you can imagine my excitement when I learned of *The Dictionary of Obscure Sorrows* by John Koenig,[1] within which he coins new English terms for emotions we feel, yet have no words for. John defines one such word[2]:

```
sonder: noun. the realization that each
random passerby is living life as vivid
and complex as your own
```

This was open source for me, and it might be for you too, now that you're reading this book. I remember peeking into my first project, observing whole histories already played out, planned, and performing in real time. Little subcultures of loosely-aligned folks working together. It was all a bit over my head...but

I *did* have a new perspective leaving that place. Once you attune your ears and eyeballs, it's hard to miss. Open source software is more than a license. Far more than *free* code. Open source software is *code* plus *community*. It's a vibrancy and a sprawl and a fulcrum and a tool and a burden and a beacon. And it needs you.

Communities whir along unseen, yet shape the products we use, the news we hear, the jobs we apply for, or the conversations we have with loved ones. Open source is all around us, omnipresent in our digital, professional, and personal lives. You'd be hard-pressed to find software that doesn't lean on open source somewhere in its composition, design, or delivery. It's a medium in and of itself. It's the tug-of-war rope via which ideological, economic, and cultural contests play out. Open source is a mirror, projecting whatever predispositions we wish onto the moment.

You already contribute to the open source ecosystem—you just might not see it that way yet. First and foremost you participate as a consumer, as a user of things. You browse and build websites atop open source software. You read documentation. You visit the repository of a project your team uses as a dependency. And so does everyone else. The company you work for, the school you attend, the businesses you visit—they run on open source software.

The Linux Foundation estimates that 90% of production software is of open source origin.[3] Researchers at Harvard Business School calculated the total demand-side value of open source at 8.8 trillion dollars.[4] It's nearly unfathomable in its scope. I think Oscar Wilde was bullish on open source way back in 1892 when he wrote this dialog for *Lady Windermere's Fan*[5]:

> *Cecil Graham: What is a cynic?*
> *Lord Darlington: A man who knows the price of everything, and the value of nothing.*

We'll cover the many simple and complex costs associated with this medium. But don't lose sight of the value too. Open source software carries value beyond what's coded and licensed. Participating in this valuation is where community forms.

You might have hesitated at the sheer size, the complexity of the tooling, or the pace of the community. It's intimidating to engage with something mature and existent. *They have a circle of friends already. I don't belong here.* But I suggest: you need not see these communities as in and out groups. Nor do you have to be a midnight developer. Participating within the open source software ecosystem can be a steadfast, rewarding exercise even when your work isn't the next blockbuster community darling.

It isn't a one-sided exchange either—playing the part of the audience is as essential to a successful performance as playwright, stagehand, or actress. We can engage with open source software along a broad spectrum of activities, activities that appeal to every skillset, experience level, and time constraint.

What is Open Source Software?

Before we get too far, I should summarize how we got here. Folks in the 70's and 80's were bartering electronics in garages and parking lots as part of movements like the Homebrew Computer Club. Hobbyists shared patches of software built atop these machines and operating systems, sometimes with dubious and uncharted legality.[6] The collective sharing spirit was strong, yet competed with corporate interests. Free software threatened these proprietary software monopolies. The GNU foundation states: "The nonfree program controls the users, and the developer controls the program; this makes the program an instrument of unjust power."[7] The message was bold, with little room for compromise. Early pioneers crafted memorable statements like "the four freedoms of free software"[8]:

> *The freedom to run the program as you wish, for any purpose (freedom 0).*

The freedom to study how the program works and change it so it does your computing as you wish (freedom 1). Access to the source code is a precondition for this.
The freedom to redistribute copies so you can help others (freedom 2).
The freedom to distribute copies of your modified versions to others (freedom 3). By doing this you can give the whole community a chance to benefit from your changes. Access to the source code is a precondition for this.

Open source spun out of this free software movement in 1998, the result of a concerted marketing effort accompanying the free release of the Netscape Navigator web browser and its source code.[9] The term *open source*, suggested by Christine Peterson, appealed to businesses balking at the idea of free software being unsellable and cheap. At its core, open source software is accessible code governed by a license—rules defining its use, reuse, and distribution. Since 1998, the Open Source Initiative maintains a formal definition of what constitutes open source, as well as which licenses adhere to this definition. The summary here is a window into a neutral, open-playing field[10]:

1. *Free Redistribution*

2. *Source Code*

3. *Derived Works*

4. *Integrity of The Author's Source Code*

5. *No Discrimination Against Persons or Groups*

6. *No Discrimination Against Fields of Endeavor*

7. *Distribution of License*

8. *License Must Not Be Specific to a Product*

9. *License Must Not Restrict Other Software*

10. *License Must Be Technology-Neutral*

Wow, that seems like a lot, doesn't it? To be honest, I hadn't even considered all those nuances until researching this book. But I've lived through some of these clauses coming into contact with communities. We'll dive into that in Chapter 4. The important thing is, folks have been thinking thoroughly about how to create maximum access to software for a good long while.

These early ideas set in motion a lot of the mindshare we work within today. Supposed truisms like "given enough eyeballs, all bugs are shallow," "contributions welcome," or "release early, release often" feel like part of the DNA of open source, even as reality is often more nuanced than described.

We take this foundation for granted now, and have transcended it with a greater emphasis on community and inclusion. Corporate interests direct and divert energy into the movement, and it seems as if we're always at the edge of crisis. We must expect more from open source and from each other. It's not *just* a license or *just* code anymore. Social coding via platforms like SourceForge, GitHub, and GitLab revolutionized access and nurtured community like nothing before them, but didn't always do it equitably or with safeguards. There is still more work to do. Mature community forms when people, tooling, and process combine to solve a problem. Lasting community forms when we have a safe place to participate and learn from one another.

Symmathesy: The Learning System

That last point is a great one to emphasize. Together, we grow and learn from one another in ways not possible alone. Our skill acquisition accelerates, our perspectives diversify, and our networks expand when we share experiences. Writer and educator Nora Bateson coined the term *symmathesy* to give this concept a single word.[11] It's derived from the Latin *sym* (together) and *mathesi* (learn). Bateson defines the word as "an entity composed by contextual mutual learning through interaction." You never have to use the word *synergy* again.

If that doesn't sound like the highest form of the human experience, I don't know what does. It's not just the individual, nor their output, but the ripple effect we have on those around us. Imagine a graph of vertices and edges representing a system. A moving web never at rest. Jessica Kerr applies the concepts of symmathesy to software development, where she describes a development team as a symmathesy, "composed of the people on the team, the running software, and all their tools."[12] The team learns from one another, users, and the software itself. The software learns as changes are made to it. The tools learn as the team turns data into information. It's brilliant stuff—she continues[13]:

> *This flow of mutual learning means the system is never the same. It is always changing, and its participants are always changing.*
> *An aggregate is the sum of its parts.*
> *A system is also a product of its relationships.*
> *A symmathesy is also powered by every past interaction.*

Open source software is a form of symmathesy. As consumers, we are direct participants in this system, drawing lines as needed. As maintainers of an open source project, we increase the size, robustness, and interconnectivity of the web. Connection is inherent in every action. The documentation, the code itself, the dependencies on other projects—the sharing is explicit. And the best projects amplify their reach via community. Participants may not even know each other, yet issue logs, code reviews, forum discussions, chat presence, and events help build a collective understanding of one another's common goals and challenges.

Individual projects can become a part of a larger ecosystem. No single person is responsible for these emergent systems, but these systems may become *reliant* on one. They're interconnected, fragile, and prone to falling over. (FIG 1.1)[14]

FIG 1.1: an xkcd comic showing how all modern infrastructure, particularly open source, is an unstable composite, propped up by volunteers. "Dependency" xkcd comic courtesy Randall Munroe.

Cascading interactivity can lead to long, confusing open source dependency implosions. Projects and people can become victims of events many levels abstracted from their application code.[15] Those aren't fun days. But this same sprawling network creates resiliency in the system too. Projects that deviate from community expectations are abandoned or forked. People lose interest, give up, move on. A variant of the original project emerges, the system reshaping itself. Sometimes those unexpected events surprise us. Community as an ever-changing web composed of us and our actions, learning together.

Act Like an Improv Artist

If you've ever gone to an improv event or seen one on television you'll understand the concept of *yes, and*.... This phrase challenges the artist to incorporate input from others and evolve their own approach in kind. Yes, and... "is a state of mind that all of the performers adhere to."[16] Embrace and extend. Acknowledge and advance. *Yes, I will use that salmon you handed me to call my boss and demand a raise.*

It's called the first rule of improv because continuing the narrative while adapting to external circumstances is essential to connecting to the audience. How well the artist accomplishes this without hesitation helps the scene. It'd be a bummer if they threw their hands up and said, "no, I don't want to break into song and dance right now, can we do something else?" Or, "no, the landlord didn't suggest we sublet with four robots." Improv masters think on their feet, build upon a shared story, and continue pace. Audience participation is the best catalyst, even if they need coaxing. Together everyone learns what the story is and where it's going, creating a community through trust and practice.

Using and maintaining software is a constant improvisation. Generals for over one hundred fifty years have said, "No plan survives contact with the enemy." I'll borrow from another general in adding that, in some cases, we have met the enemy and they are us. Turns out, we're wildly inconsistent as a species, let alone collaborative software endeavors. The slickest project plan in the world cannot account for the speed of open source.

What is the speed of open source? It's both imperceptibly slow and breakneck fast. Every day a social media feed teases something new. Must you stop to learn it? *Am I missing out?* Something is broken and the status quo is at risk. Take dependencies. It could be that `0.X` version of a library that hasn't been updated in three years, suddenly releasing a compromised version due to missing two-factor authentication on the maintainer's npm account. It could be a maintainer forgetting a file and publishing a broken release. Or it could be a well-known project unwittingly releasing a breaking change during a routine update. When these things happen, often there are high-profile herd-like behavior shifts to other packages, workarounds, reversions. Oh, and plenty of angst.[17] Someone might propose a solution and the maintainers need to quickly assess the risk of merging that change. It might be against current patterns, lack tests, or add yet another configuration option.

As bystanders, we need to assess how to move forward with our own consumption plan. How we respond to those moments is key to the long-term health of the project. As a group we work to learn and apply new knowledge in this ever-changing environment.

Seek Reciprocity In Relationships

We don't work in silos. Even as individual contributors, freelancers, or lonely teams of one, we interact with other people, or their facsimile products, throughout our workflow. You might be designing a site for your partner's business. They will review, interact with, and be affected by the impact of the site on their end users. Making the site useful, accessible, performant, and maintainable helps all parties continue to engage with one another. A deficiency in any other of these areas imbalances relationships. Often the user is the victim of these imbalances, and they may leave as a result. You miss out on user flows, feedback, or revenue.

Ask yourself, "what is the expectation I am suggesting?" If you wish your project to go beyond "be glad it exists," you must *design* your users' experience. The lack of this design is a choice as well, but not a very responsible one. What does someone get for investing their time in learning more about your project? Are you respecting their time as much as yours? Are you setting them up for success or failure? The absence of this balanced equation between parties creates incongruent relationships.

Irresponsible sharing of maintenance can lead to atrophy, risk, and wasted effort. Our intentions are not enough. Think of this as the squandering of a public resource. *Someone else will pick up that litter. That dude cut in line, so I can too. I don't need to report the issue, or follow the issue template.* Collective inaction, missing governance, or everyone expecting others to pitch in can lead to erosion of a once-firm footing. Find ways to over-communicate your intent or your understanding so that assumptions are lessened and agency is clearer.

Your intentions must be defined and implemented via specific policy. Park staff can place garbage cans all over the picnic area to encourage responsible disposal. But then they need to pay someone to empty them, and they attract bugs and pests. Or they can be bold and have no garbage cans; instead installing signs to "leave no trace" and to haul away refuse as you leave. Radically different approaches rooted in the same goals. The expectations are clear in both policies—the maintainer communicates; the consumer must respond. Lots of open source projects suggest that you open an issue clarifying work and claiming it as your own before beginning development. This gets everyone on the same page before committing a single line of code. Others may tell you not to "lick the cookie,"[18] or claim a task with best intentions but never complete it. Use the contribution guidelines when considering contribution. Observe and emulate patterns of behavior when approaching a new community. Don't litter, and don't pick the wildflowers.

It's too easy to fall into this familiar consumption pattern with open source software. We take but never give. We are busy. Too many issues to wade through for a solution. Another `npm install` doesn't hurt, does it? Well, aggregate web page analysis[19] shows broken budgets across size and request count metrics. As consumers of open source software, we must be aware of our part to play in perpetuating incongruent relationships. We must likewise consider the people creating open source software. A download isn't enough. Waiting for someone else to report it or fix it isn't enough. Donating may not even be enough. Maintainers crave our *time*[20] (mentioned 12 times in a single comment by the Moment.js maintainers). We need to invest in the projects we use via thoughtful engagement.

Denis Pushkarev highlighted the incongruity he experiences as maintainer of the core-js polyfill library. He penned a lengthy (a quarter the *entire draft* of this book) accounting of the imbalanced relationship the world's largest websites have with his project, built to shim old and new web features to the broadest set of browsers. By his measure, his code is used by over fifty percent of the top ten thousand websites and seventy-five or eighty of the top hundred.[21] According to Pushkarev, that's an install-base only surpassed by jQuery. He doesn't point this out looking for kudos, "but to show how bad everything is." At the time of writing, Pushkarev claimed full-time development amounts to about four dollars an hour. Atop all of this, he calls out the targeted abuse he routinely receives because of his efforts to earn more. His is a common perspective, and informed from experience, not social media punditry. Anyone would question their resolve with these factors arrayed against them.

We must do better. We must rebalance relationships as consumer and producer.

Observed Systems Are Changed Systems

Complicated things are better understood when taking them apart. The totality of the machine may be hard to grasp at first. But gradually, with both study and practice, the complete picture is rendered in your mind. Other times you surprise yourself with a discovered inefficiency—something that can be added or removed to improve the whole.

Outside scrutiny will most certainly impact the mood, trajectory, and lifespan of a project. Public code simultaneously informs and is informed. Symmathesy. Remember, an audience has as much sway and investment in a successful performance as the players. Knowing that you are being observed yields five to twenty percent scientifically measurable[22] improvements in performance too. That's why *shipped* is better than *perfect*, or why adopting the improv artists' strategy of *yes, and...* encourages a growth mindset. You'll be better prepared to contribute to a project if you are in close proximity to it. Sit in the front row. Clone the project. Take it apart. Break the tests. Create a mental model of its uses and constraints as a composition arrayed against other factors. Formulate an opinion about the software and then see if it holds up to the opinions of peers.

Oftentimes the mere act of using software will reveal opportunities to improve it. A foundational python style guide states, “code is read much more often than it is written.”[23] And once written, a maintainer may skim, internalize, or ignore their own documentation. The humble typo is a psoterchild of this behavior. The fresh eyes of someone reading the code for the first time, from a different context, will find something a seasoned or original contributor may miss. This naivety is open source gold—spend it generously and hope your money is good where you choose to send it. Maintainers not listening are not worth your time. Maintainers of mature projects will create pathways to contribute documentation or code comment fixes with as little friction as possible. As a maintainer, encourage raw, inquisitive, and “stupid” questions. Under their surface is usually an unmet need or an assumption to be rooted out in the code or docs. For every person that comes to you with a question like this, imagine dozens more getting stuck and saying nothing. These folks using your project are choosing to engage with you—I promise that’s a big deal.

Abstractions

I’ve written this book as a new entry in the compendium of open source thought. I don’t fancy myself a thought-leader and for that we are all fortunate. But it is informed by years of success and failure working with open source software. The advice is honest; the vision bold; the mandate modest. I was once hesitant to understand where I fit in the open source community. I suspect some of you might feel the same. I found footing in a project called Pattern Lab,[24] within which designers and developers could create larger and larger components using atomic design patterns. I enthusiastically used, contributed to, forked, maintained... and burned out on it. It was as frustrating as it was rewarding. It taught me what I hope to teach you. Open source brought about the best in me, not because of any particular commit or feature, but because of the vehicle it was to further my connection to others. I hope that sounds as exciting to you as it still does to me.

We constantly swim in the wake of others, creating our own as we go. We are both consumers and producers of what Brad Frost calls, "creative exhaust."[25] He says, "It's not about the work you do, but rather what that work enables others to do." We work together and apart toward goals, never quite seeing the fullness of the web. We sprawl and simplify. We diverge and agree. We construct and abstract. Daily life is an abstraction: from wheel to automobile, from lightning to lithium battery, from rice to risotto. We build upon the work of one another, combining ingredients, remixing content, drawing new connections. We are part of a symmathesy called humanity, or maybe jQuery, depending on the grandiosity of your focus. We push and push against boundaries until they give way, if only a tiny bit at a time. Open source is like this in its best incarnations. Keep pushing. Find your community. The remainder of this book will show you how.

Chapter Two

The Spectrum of Engagement

Many hands make light work...too many cooks in the kitchen...there's plenty of dueling idioms to describe collective behavior. As our community increases in size, so too can our motivations clash and our individual skillsets complement. If you caught the tension between these phrases, good! Open source software development exposes us to this paradoxical environment. We each seek value by engaging within our capacity. And we do that in a surprising number of ways. I define them as *consume*, *contribute*, *maintain*, and *empower*. I want to show you that there is so much more to open source than "just"[26] using it.

Value Exchanges

My then ten-year old son Max, when learning about this writing effort, said "I have one suggestion as you write your book, use a few analogies." Last chapter had plenty of those pretty overtly positioned to align with open source software—though I posit a last one (okay probably not the last): gardening. It's not a novel analogy, as others like Maggie Appleton use it effectively to organize content,[27] but it furthers our narrative—if only we dig in a bit.

A garden is an intentional arrangement of plants. We shape the earth, enclosing and manipulating it for a purpose—cultivation, ornamentation, soil-retention, nature-preservation, research, or otherwise. It can vary in size, site, crop, focus, and care requirements. We derive different values from each garden, depending on the motives and needs of our context. Software comes in all configurations too—personal, proprietary, open source, small, large, weed-ridden. Let's focus on that open source one. Maybe you see where this is going.

FIG 2.1: A community garden is the intersection of many needs and skillsets, each as important to the success of the effort. Photo by Steve Adams on Unsplash.

Envision a community garden (FIG 2.1)[28]: now there's some fertile territory for analogy. Max will be proud. Many actors combine efforts, motivations, and needs within a singular plot.

This garden requires planning across every dimension. You need physical space. Is a location secured? Is it near people that can work and benefit from the yield? What will the layout of the space be? How will plots be divvied up? You need raw material. How do you pay for the soil, tools, seeds, water, and fertilizer? Do you need a fenced enclosure to keep out deer and rabbits? What will you grow and how will it not die because everyone else thought the other folks would water and weed once school started and it got busy?

A critical point to be made here is that you need not know all these answers to participate. Maybe the spark of engagement is accepting some free, surplus produce from this year's crop. Maybe you start with gumption and a trowel. Or nothing at all but your attention and time. By gardening a bit yourself, or if you've grown at the community garden for a couple seasons, you'll accrue general and specific expertise that makes you more and more comfortable taking on tasks. Through proximity and persistence, you might one day find yourself managing the fundraising campaign, mending the fences, bringing the produce to a local shelter, recruiting neighbors to pitch in, or assigning the seasonal plots to keep the Montagues and Capulets separated from one another.

The sowing of seeds, mulching, watering, weeding, harvesting—you know, the actual gardening—that might take a back seat to the needs of the moment or your shifting interests. And that's okay! Because at this scale no one person can do it alone. Rather, it takes a group of committed individuals—a literal community—working together to make the effort succeed. The array of activities we choose forms a spectrum. We might not undertake tasks along the spectrum all at once, in sequence, or at all. Tasks may change in complexity, responsibility, or time-investment. We generate different values from each of these actions, each an investment in the whole.

All of this takes time. Gardens are not transactional—they are a relationship, a long-term investment without the promise of success at the outset. The time and care needed to nurture a seed through to harvest requires more than a moment or payment. Of course, you can buy a carrot for less money than your time is worth growing one from seed. But that's not really the point. Instead, the payoff—the value—is both the process and the product. Working in a garden has an implicit value *exchange*: effort and resources over time create value. Open source software value maps to this same exchange. Time is our currency, whether that's:

- **conserved** in using someone else's code
- **wasted** as a result of poor implementation, bugs, breaking changes
- **invested** via contributing fixes and features
- **donated** to others through maintaining

A transaction is momentary. What open source projects create are natural opportunities for lasting association. And the best part is, we can learn to think of these actions along a continuum wherein we can play the role or roles we want.

The Spectrum of Engagement

By consuming open source software during your day, you've already engaged within the open source ecosystem. But there is a more complete picture, defining a spectrum with distinct activities for different roles. A broad array of options lay before you, like all those tasks in the community garden. They all need doing and have merit in advancing project goals.

At one (or many moments), you may *consume*, *contribute*, *maintain*, and *empower*. This is the *spectrum of engagement*. (FIG 2.2)

FIG 2.2: The spectrum of engagement covers a range of activities that define how you can interact with a project.

This isn't some abstract software concept. I acknowledge the cringe associated with business-speak like *engagement* and *value exchanges* and push past them. We can reclaim their humanity. Stay with me. One can overlay the spectrum of engagement atop most aspects of life and see the range of community-driven action. We are at all times exposed to, and, if able, participatory in community. Even if you don't directly act, you are affected. More on that in Chapter 6. Let's apply the spectrum of engagement outside the realm of open source or even tech. Think of another intersection of actors and activities: accessibility in businesses that are open to the public. The goal is equitable access to goods and services.

- Title III of the Americans with Disabilities Act[29] outlines the criteria and requirements businesses must comply with.
- Not only do persons with disabilities benefit from increased access, but everyone according to their situation and context.[30]
- The communities they participate in and businesses they patron benefit from their presence.
- Engineering firms and architects comply with federal standards when drawing new plans. They factor in braille signage and accessible switches, hooks, door handles, grab bars, dispensers, driers, and restroom fixtures. They design adequate flat maneuvering at the top and bottom of ramps for motorized scooters, wheel chairs, and service animals.
- Organizations contribute to these outcomes with direct advoca-

cy. The Barrier Free Golf project in Chaska, Minnesota designed a first-of-its-kind golf course promising "people with disabilities, veterans, juniors, and beginners of all ages access to fun golf."[31] The Ramp Up Iceland project[32] organizes construction of access ramps across the country.

- Litigation in pursuit of equity bands interests together toward common goals too, such as when the City of Philadelphia settled in court to upgrade ten thousand curb ramps around the city.[33]

Maintaining these efforts requires investment, regulation, and willpower. Empowering change in the first place sometimes requires a community of citizens or concerted lawmakers, but always benefits from an engaged populace.

The spectrum of engagement applied to open source software allows you to dive in across a range of activities. (FIG 2.3) It's user-centric and not necessarily oriented toward growth from a maintainer's perspective. You don't have to do the "lower" tasks before leveling up to the "higher" ones. I suggest they are all of value to a mature project. Growth in role happens, but it's not the outright goal.

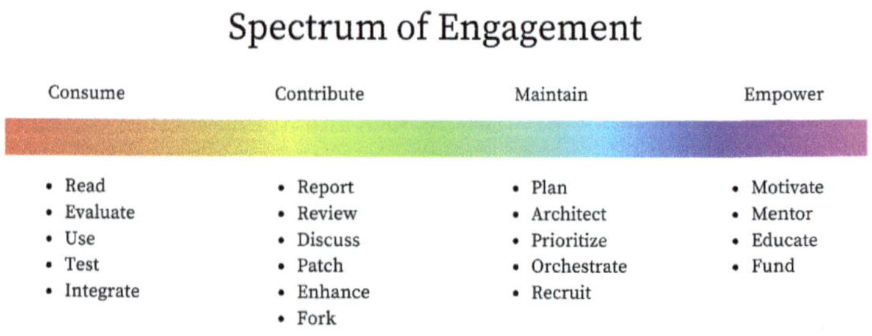

FIG 2.3: The spectrum of engagement contains many tasks that further the goals of a project.

Your familiarity with an open source project will increase from using it. One day, reading the docs for a tool brings you to their issue log after you notice a small typo. This is a moment of action, if only you pause to take it! You submit the fix for the typo, a small but engaging moment. You usher the patch through their pull request and deployment process, learning a technique or two along the way. You've converted a bit of your time into advancing both the project and your knowledge. You didn't set out to "do open source," as Salma Alam-Naylor suggests[34]:

> *Asking "How do I get started with web dev?" is like asking "How do I get started with cooking?" Decide what you want to make. And then reframe the question.*

In your case, you decided to improve a specific thing and went from there.

All parties benefit from the spectrum of engagement. We all want value from our efforts, even if we don't consciously think about it. How is our time and attention balanced out on the ledger? Everyone wants expectations set and met for their investment. As users, we can be vocal in support and constructive in critique. As maintainers, we can be clear in values, priority, and polish. We can make pathways to increase engagement. We all, as participating citizens in this ecosystem, can affect the most change together through discussion, collaboration, and listening. Our roles may change per project, per ticket, per day. Let's examine these roles and their activities, engaging along the spectrum.

Consume

Sort of like hosting in "the cloud" means hosting via "other people's computers," consuming open source software means using "someone else's code." Kelsey Hightower expands on this trope, revealing the depth and specialization behind it all:

> *The cloud is just someone else's computer, power, cooling system, underwater sea cables, networking equipment, data center and the land it sits on, insurance policies, roadmaps, and a whole lot of people to build, secure, monitor, and support all of the software that runs on top.*

We'd be mistaken to think of the cloud or open source software so shallowly and without regard for the humans on both ends. This is where the value exchange between user and maintainer can feel most imbalanced along the spectrum. We may feel guilt or imposter syndrome consuming someone else's code rather than writing our own. *After all, if I don't parse and localize those dates from the database myself, who am I as a developer and person!?* Fear not! I challenge you to feel relieved, resourceful even. This may actually be a form of maturity—one often remixed within memes highlighting that both newbies and experts enjoy simplicity—yet a sizeable group of folks work harder in pursuit of intellectual purity. (FIG 2.4)[35]

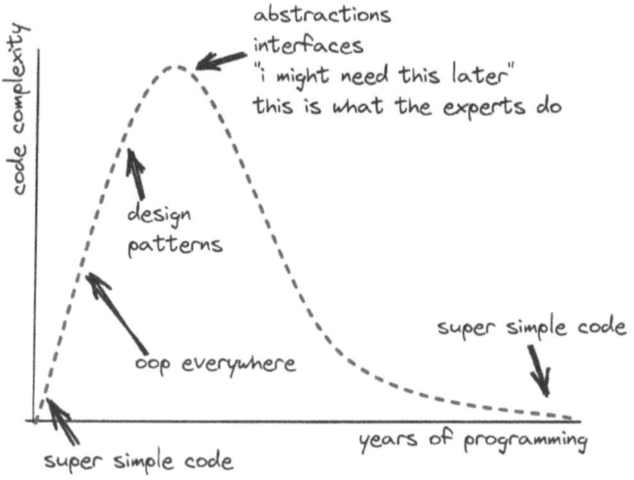

FIG 2.4: A bell curve of programming effort expended as someone gains experience. Beginner and expert alike chose simple solutions while those with mid-level experience work harder at dubiously useful endeavors or optimization. Image by Flavio Copes.

Consuming open source software exposes and expands your skillset and approach. It is a form of risk; a calculation that, if conscious, can keep you moving. We'll talk more about this tightrope walk next chapter. As you work in this space, you'll notice organic opportunities not only to learn but also to share what you've learned. This is your superpower. You will formulate opinions, perceptions, and maybe even expertise in the tools that you use—don't keep it to yourself! The work you've done leaves callouses and wisdom. In what ways can we consume open source software? I think you'll be surprised at the array of activities before you. Let's break it down.

Read

Learning about a particular piece of software increases your awareness, draws connections based on existing context, and helps you align it against the mental model of potential utility. I remember reading of jQuery's release (over ten years ago) of a new API that allowed teams to level up from direct to delegated event handlers.[36] Rather than attach event listeners to the element taking an action, like a button in each table row, we could delegate the action upward to the table itself. (FIG 2.5)

```
// attaches an event handler to every button in the table
$('#dataTable .js-btn-edit').on('click', function (e) {
  console.log(e.target.innerText)
})
//.attaches a single event handler to the table, listening for clicks from the
child buttons
$('#dataTable').on('click', '.js-btn-edit', function (e) {
  console.log(e.target.innerText)
})
```

FIG 2.5: Optimized event handler registration made possible with a newer version of jQuery.

The functionality is the same and the syntax similar, but the performance impact compounds the more rows you add. (FIG 2.6) More importantly, the `on` delegation method also handles future items being added to the DOM, eliminating entire classes of concerns for web developers. For example, adding items to a list or table wouldn't require you to add more event handlers. This small change improves flexibility, code organization, and readability. Reading about this only clicked when I could piece together jQuery in relation to a browser's native event model.[37]

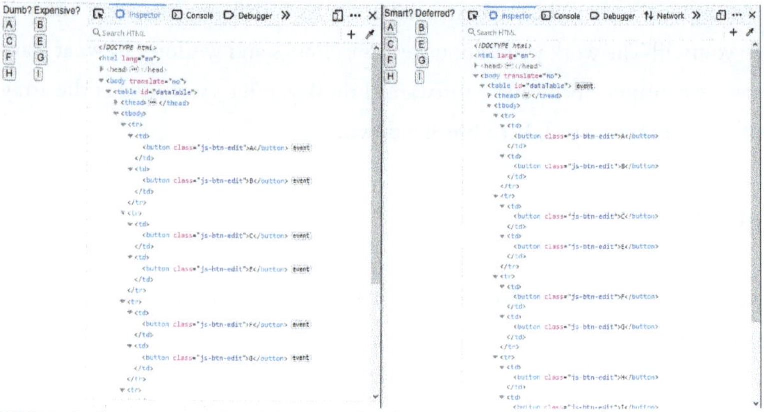

FIG 2.6: jQuery examples of direct and delegated event handling. On the left is the inefficient event handling strategy, with a handler on every button. On the right is the efficient, delegated event handling strategy, with only a single handler on the table.

I was more prepared to assess event binding scenarios in my work and others'. My novice eyes opened to the idea of maintainability and algorithmic efficiency within a website and how developer decisions could directly affect DOM performance and end-user experience. We must educate ourselves to better weigh tradeoffs like this, whether it's jQuery API choices, CSS pre-processor bloat, or React layout thrashing.

Evaluate

Comparing one piece of software to another, or against criteria, helps you discern what matters most. Objective analysis is not always possible, but grounding your evaluation against known constraints can help sift through signal and noise. Busting out a weighted decision matrix in Excel is a great way to make friends at parties, and it also helps you slow down and apply reasoned rubric of thought to a situation. I've already had to defend evaluation decisions about software made many years back. Having a thorough evaluation procedure, data, and stakeholder buy-in helps maintain alignment. It also gives you headroom should things not work out. We'll cover all this in more detail as it applies to dependencies in the next chapter. Isolate to understand how a solution works before comparing or integrating it. With content delivery networks (CDNs) hosting libraries and online editors, it's never been easier to play with a project for little-to-no investment. Comparing and contrasting solutions prepares you to speak up or have opinions about one piece of software over another.

Use

Exercising the API is a surefire way to better understand the software, if only at a surface level. I've used that acronym a couple times now and it would help to define it. API stands for "application programming interface." You can think of this as the public-facing featureset, be it a web service, a library, or maybe even a product. In a more technical sense, an API is the contract the software makes with its users. The API for a calculator has an addition function that says, "you give me two or more numbers, I add them together and return the resulting sum." You will inevitably find a missing feature, deficiency, edge case, bug, or oversight when using software. This can lead you down many paths of engagement. Do you look at the documentation? The source code? The issue log?

Test

As the amount of software you weave together increases, so too may your desire to have confidence in those bonds. And we don't want to test someone else's open source code—*they have their own tests, right?*—so what can we do? Testing your own software alongside open source code often leads to the need to mock out APIs. Mocking, especially within unit tests, limits the scope of your test to the code that you wrote. Many testing tools provide a means to mock third-party dependencies. Here's the Jest library testing a function called `fetchList` and mocking the axios network library. (FIG 2.7) We mock the `axios` module to always error, and return a rejected `Promise`. The unseen `fetchList` implementation catches this error and returns an empty array.

We learn a lot about axios's API by testing it, such as its `get` function that returns a `Promise`.

```
import axios from 'axios'
import fetchList from './myDataService'
jest.mock('axios')
describe('myDataService', () => {
  it('returns an empty list on error', async () => {
    // arrange
    axios.get.mockRejectedValueOnce(new Error('oh no!'))
    // act
    const result = await fetchList()
    // assert
    expect(axios.get).toHaveBeenCalled()
    expect(result).toEqual([])
  })
})
```

FIG 2.7: A mocked instance of the axios dependency, which we learn a lot from by testing its API.

Mocking may eliminate the need to make a real network call, or replace specific time-based values with a static date and time. Doing this is a great way to get a feel for the input / output of an open source project, as you need to emulate the API of their libraries in order for your tests to use facades of the real thing.

Peeking at a project's tests can often be instructive too. Mature projects will have exhaustive test coverage exercising their own API across different permutations of architecture, runtime, configuration, and language. I've seen hidden features, command-line options, and parameters only documented within tests. Tests may contain usage hints or constraints that can help you expand your opinion of a project. You may encounter a scenario or edge case a maintainer has not accounted for within their test suite; what a great opportunity to engage!

Integrate

Similar to testing, composing open source software into larger and larger applications, or applying them to solve new use cases, can increase your knowledge of the domain and toolset. This prepares you to engage and apply your learnings to other scenarios. Understanding where the software fits into a bigger machine forces you to work within its API boundaries, read its documentation, and think through how systems interact with one another. This consumption can be creative, divergent, and chaotic. I remember people doing things with Pattern Lab I had never dreamed of needing, let alone trying. My mental model didn't align with their needs and had to expand. When those folks engaged with the project, they pushed the envelope, introduced a new capability, assuaged a concern, or made the project more approachable. We all had an opportunity to make it better.

As you use open source software, you might attune yourself to a vision beyond what's available. It's time to contribute.

Contribute

If and when the time is right, you might choose to do more. Active participation on a project can feel exhilarating. I remember the excited moments making my first pull request to picturefill,[38] Scott Jehl's JavaScript polyfill library imitating the proposed spec of the new `<picture>` HTML element.[39] Discovering new tech being built in the open, and being in a position to help build it? It opened my mind to concerns broader than my current work, role, and skillset, all in a quite humble set of code changes. Connecting with other people around the world interested in the same things as me created comfort and energy. We as a community were in open dialog with each other as developers, specification authors, and browsers vendors. The results mattered, but equally so did the side-effects: symmathesy was occurring around responsive web design, image performance, and user experience.

The longest-lived open source projects tend to evolve tiered contribution models. This hierarchy is a product of size and conservatism. They may appear to be gatekeeping, such as the structure the Apache Software Foundation assigns to its projects' roles.[40] User → Developer → Committer → Project Management Committee Member → Project Management Committee Chair. Each role feeds into the next and has unique responsibilities. But beyond the structure, one realizes this is the spectrum of engagement at work again—participation begets increasing familiarity and responsibility within a project. The Node.js project uses a nomination process that highlights a potential collaborator's total contributions, reviews, comments, and created issues,[41] not just code pushed. This holistic model of contribution better recognizes the value exchanges inherent in the spectrum of engagement.

Report

Participating in an issue log is fertile ground for learning. They also need weeding, like any good garden. In the Node.js project, one of the better places new contributors can spend time is in triaging issues, "enabling newcomers another meaningful way to get engaged and contribute."[42] Some projects like lodash and Eleventy use a system where all feature requests are closed in the issue log but then voted upon by users. From a quick sort it's easy to see what the community wishes, and how many requests need to be prioritized against each other. (FIG 2.8)[43]

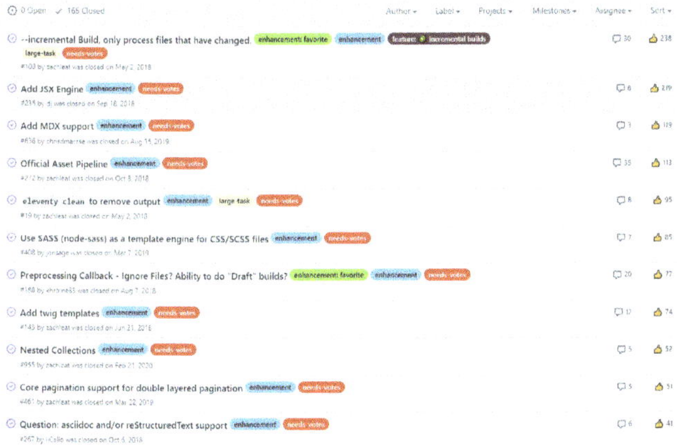

FIG 2.8: Eleventy's issue log wherein issues can be voted on by the community for prioritization.

Connecting with others encountering the same issue furthers your comfort with the project. Reading others' issues builds empathy, and challenges or deepens your understanding. Corroborating an existing issue with more evidence helps maintainers prioritize work.

Speaking of maintainers, writing your own issues forces you to think like a maintainer. What is someone supposed to do with the information in your user report? Is it actionable? What can you provide to better compel them or others to action? A well-structured issue report is integral for time-starved maintainers to make sense of your problem, offer help, or suggest next steps. Take the time to review the project's contribution guidelines and adhere to their issue template if present. Someone spent time to research how best to help you so the least you can do is listen to it.

Issue logs are a melting pot of cultures, skillsets, expectations, and languages. Assume positive intent here, but proceed at your own pace and comfort. They can be the public bathroom wall of open source, a place to scrawl grievances with little concern for repercussion or impact to the recipient. Try to frame your interactions as if you were talking in-person to a colleague. Engage only when you have the motivation and energy to do so.

Review

I saw Dan Cederholm deliver his *Handcrafted Patterns* talk at An Event Apart Atlanta 2012.[44] What stuck with me from his talk was the ways in which we can learn from one another. Luke Wroblewski was also there and took better notes than me[45]:

> *There's a common progression to learning anything creative: imitation, repetition, and finally innovation. Imitation: by doing what others have done. Repetition: learning and repeating patterns. At a certain point, muscle memory kicks in and you can use patterns without thinking. Innovation: apply your own creativity to the things you've learned. This adds your own style to things.*

Seeing someone else do nearly anything, well or otherwise, can directly inform our own methods. Code-reviewing someone else's work, or just reading it, can reveal novel approaches, new tools, and different thoughts. And, you might be able to do the same for them! If you are looking to help out within a mature or emerging project, but don't know where to start, helping oomph in-progress work over the finish line is an admirable endeavor.

Discuss

Forums provide longer-form topics or dimensions parallel to the code itself. Some projects use discussions to announce new releases, or intentions to build new features as an RFC (request for comment). Message boards of any kind are a public means to rapidly build context necessary to make more effective contributions. Engaging in the discussions makes you better informed on the project, helps set direction, and builds your relationships with others using the solution.

Patch

A patch is the gateway into larger forms of contribution. As mentioned, a typo or docs clarification is a great way to start this engagement. Tools don't even require you to leave the browser, offering text editor capabilities per file or full-featured development environments. (FIG 2.9)[46]

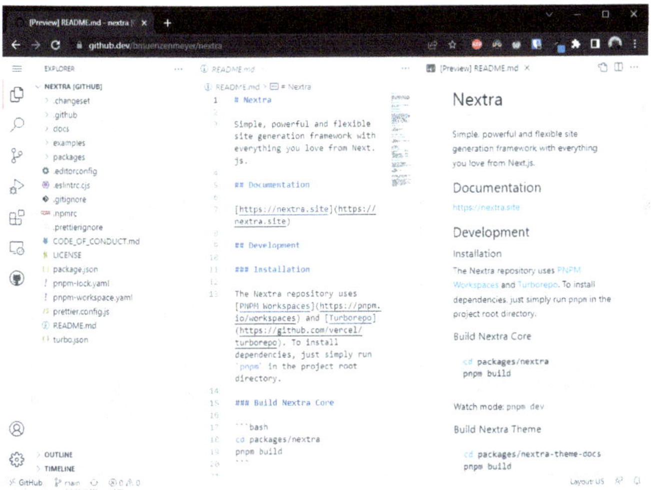

FIG 2.9: A repository loaded within GitHub's online editor. A GitHub codespace can further enhance this with a dedicated virtual machine, all within the browser. These features lower the barrier for entry for contribution.

The longer you use a project the more likely it is you will find something wrong with it. This is an opportunity knocking, don't ignore it! Oftentimes a bug fix will expose you to the test suite, or the pull request process for the first time.

Enhance

Open source solutions often lack robustness in every use case or constraint. The projects "move at the speed of open source" or adhere to the notion of "squeaky wheel gets the grease." And why wouldn't they, really? We should only take what we need at the outset of a journey. Enhancements you identify or identify with can be directly contributed to the project—a ramping up of engagement that usually is a larger investment in time and code. Look again to an issue log for "good first issues" or items within your interest and skillset. Make meaningful contributions that advance the project forward, even incrementally. Familiarize yourself with contribution guidelines and the release process. Recurring work here, whether it's feature work, fixes, updates, etc. is *maintaining*, which is a whole dimension that we'll cover in earnest.

Fork

Within GitHub, forking is a direct copy-paste of the code within a project into another owner's space. Many projects encourage contributors to work within their own version of the code, and then contribute back changes as needed. You might also find projects where you are tempted or outright encouraged to fork them. Forking allows for an inheritance of sorts of project code, goals, problems, and even community. Users within restrictive enterprise environments might resort to this as their only means of iteration. Decisions need to be made about whether or not to maintain a relationship with the upstream project once you have forked it. All in all, forking could serve a momentary purpose and then be deleted. Or it could become a sustained presence in the ecosystem, like it did when I forked Dave Olsen's PHP version of Pattern Lab into the more comfortable and ergonomic-to-me Node.js version.

These are all great options if you'll looking to give a bit back to the projects you rely on. Who knows, maybe you'll become a regular.

Maintain

Someone that's consistently acting in the long-term best interests of a project is a maintainer. It's right in the name—they are invested in the project strongly enough to make sure it's cared for. Maintainers can arrive at this point through two primary means:

- Original maintainers building the project from scratch
- Sustained contributors leveling up into long-term stewardship

However maintainers arrive, they must take great care to find balance. It's naïve to think that any one person is an expert in all the dimensions that make a project successful, can do all of it at once, for a sustained period of time, and to a high-degree of satisfaction or effect. That's why attracting a big tent of varied talent helps us assemble our open source Voltrons. Later chapters in this book dive deeper into maintenance concepts, so here we'll only outline work significantly different than contributing. I sometimes see this rubber-stamped *governance*, but with little context. Let's break it down into internal and external activities.

Plan, Architect, and Prioritize

Within any project, competing needs are inbound and released software should be outbound. Maintainers are air traffic controllers trying to land user, community, tech, and personal priorities. Without this work, we don't have a project—we have a pile of code. Just some of the tasks in need of equilibrium include:

- **Planning:** You know, the "Who, what, where, when, why, and how" of any work. Keep a backlog well-maintained so you and contributors know what to work on next. The more detail, the better context for others to build from. But be careful not to overdo it—a logjam of good but incomplete and unprioritized ideas is almost worse than nothing. A good rule is if you don't see getting something done in the next six months, delete it. If it's important, it will come up again. Beyond six months, everything will have changed anyway—user needs, the technical approach, etc.
- **Architecting**: Adopt strong opinions about the foundations of the project. Define a set of principles and tech that realizes them. Be

scrupulous about the encroachment of duplicate techniques dueling within corners of the codebase. This happens a lot with time and piecemeal contribution. It's the maintainers' job to maintain cohesion. Ideally, write down your technical stances so you can defend them, others can internalize them, and all can evolve when presented with new info.

- **Prioritizing**: Balance personal dimensions. Set boundaries of the work into your personal time. Calculate the trade-offs between decisions to tune the value exchanges your project offers. Know when to say "no" to a request or encourage engagement in the face of complaint. Think through proper succession planning should you hand over the project to others in the community.

Orchestrate and Recruit

There's more to a software product than writing it. Tasks mundane, mandatory, or advisory all compete with individual contribution. A shift can occur with long-term maintenance, in that you code less, review more, and gradually find yourself more conductor than musician. Wow, the opportunity for another labored analogy is right there, but this time I'll give it a rest.

In our efforts to sustain momentum on a project, we again need to balance:

- **User needs and safety:** Ensure the user of the software is not met with harm. Make the product stable, accessible, useful, performant, and adherent to data privacy laws. Keep an eye out for developer convenience running opposite user needs.

- **Day to day:** Shepherding contributions over the finish lines takes time-intensive review, written feedback, and testing. Many a pull request can get bogged down by something as simple as a merge conflict or broken test, but often a contributor won't have the context of a maintainer to understand how to resolve it. Giving every little thing attention really adds up.

- **Community expectations**: Ensure the community remains healthy and engaged. Set expectations and plan to moderate deviations. For example, you might need to actively model constructive criticism and on-topic discussion. Creating an inclusive space for people to work is a prerequisite to the work. Increasing avenues to contribute to your project along the spectrum of engagement helps its long-term resilience and relevancy.

- **Growth and support**: Engage with your users to understand their goals and constraints. Approach potential contributors to gauge their motivation or self-efficacy. An enthusiastic newbie may be far more open to learn than a jaded or busy veteran. It takes time to nurture any new contributor, and it's an investment with the risk of loss. Contributors aren't a flock of sheep to collect and tend to—rather they are audience and actors at the improv show looking for cues.

These tasks are project-oriented investments. There's another role, however, if your interests are more generalized.

Empower

Advocacy and enablement stand apart from consuming and producing open source software. Empowering others to engage in the spectrum of engagement makes for stronger, more inclusive communities. Within these communities we need space; and we can create space. The space we *need* is tangible and comes from others, like granting employees permission to work on open source projects (more on that later). The space we *create* might be a workshop we organize to open source an internal tool. Whatever the action, empowering others is multiplicative magic. It snowballs into far further outcomes than any individual contributor or otherwise "rockstar" developer can accomplish alone. So how can you empower others within the spectrum of engagement?

Motivate

Incentivizing engagement can sometimes be as simple as sending word to those around you. Ask. Tell them that you need help and want them to get involved. I only got involved with my company's open source program office because someone had the wherewithal to identify my nascent interest and the initiative to ask. *Thank you, Kristin, by the way!*

Organizations, conferences, and businesses sometimes sponsor open source community events. All Things Open[47] and the Linux Foundation's Open Source Summit[48] host presentations and day-long events intended to foster community growth and energy. They funnel attendees into advocacy and outreach topics specific to open source development.

The Hacktoberfest[49] event every October encourages new and seasoned coders alike to find ways to give back to open source projects. Motivations here are usually pure—but so is my desire to eat less ice cream. The fact is, many projects are not staffed or oriented to handle the influx of new contributors. Overeager newcomers may overwhelm in their attempts to help, submitting poor quality changes like grammar semantics or arbitrary text edits. AI-generated contributions are suddenly more noise. Still, the event sponsors are going out of their way to improve our communities and open diversified "low or no-code" contribution paths that look a lot like the spectrum of engagement. I love these emphasized values upfront on their guidelines[50]:

> *Everyone is welcome.*
> *Quantity is fun, quality is key.*
> *Short-term action, long-term impact.*

We can motivate in smaller ways too. Celebrate open source projects you use, trumpeting their success on social media or via thank you's across their communities. Gestures of goodwill don't always need to be grand for them to be impactful. A personal note may be worth more to some than any monetary reward.

Mentor

Continuing the personal theme, if you have the opportunity to mentor, do so. You will not be prepared—which is a good thing. I guarantee you will learn something along the way. Teaching others always forces an additional level of mastery in a skill—even if you feel like a teacher reading one chapter ahead of the class at times. The SSH key roadblocks that seemed insurmountable the first time you encountered them now represent a thirty-second diversion and a teachable moment. Helping others open source their work or contribute to another project boosts their introduction to the spectrum of engagement.

I had the privilege of mentoring attendees to Grace Hopper Celebration's Open Source Day[51] in September of 2022. Organizers primed dozens of projects and solicited for mentors to be on-hand virtually to support conference-goers in project setup, Git configuration, coding, and pull request submission. There were layers of responsibility between organizers, project leads, and mentors that resulted in a healthy stream of activity. By the end of the day I helped over a dozen attendees set up Node.js locally and work to make their first open source contribution to that project. It was rewarding to see them progress, and I learned more about how Node.js and the team operates. I found areas of concern while helping out and contributed back fixes of my own. The whole day was a formative example that engaging in even the largest projects is approachable to anyone, and it can be made easier via designed engagement. In 2023, I found avenues to prepare nodejs.org redesign[52] work for attendees even though I had a conflict the day of the event. Participation along the spectrum of engagement felt palpable and fluid, depending on the moment.

The Outreachy organization[53] aims to connect under-represented individuals with open source projects via paid internships. Web-adjacent projects Outreachy has hosted include Apache, Firefox, Git, Wikimedia, Yarn, and more.

Mentors devote five hours a week to help interns with development tasks. Other volunteers are positioned to help answer questions and provide support. The structure of the program is well-defined and has many safeguards in place. The All In project[54] similarly brings together industry partners and organizations with historically black colleges and universities to "create more inclusive open source communities for current and future developers from underrepresented backgrounds and regions." It does this through intentional trainings, accessibility audits, language audits, and more. The value exchanges here are researched and equitable. Increasing diversity in open source only improves open source.

Large open source software foundations advance the field and create managed opportunities to engage. They actively court and mentor new projects under their wing, creating structure from an otherwise unruly ecosystem. The Linux Foundation, the Apache Software Foundation, the Free Software Foundation, the Cloud Native Computing Foundation, and the Mozilla Foundation, to name a few, provide broad support to their member projects, contributors, and users. They also act as a neutral party, shielding communities from the impact of corporate mergers, IPOs, layoffs, strategic pivots, and billionaire-fueled ego trips. This structure acts as an essential backstop on the open source playing field. We want more United Nations and less Batman.

Educate

Just like mentoring, creating learning content is an empowering act. Katie Sylor-Miller maintains Oh Shit! Git!,[55] a community site so impactful to our shared understanding of version control fiascos that it has been translated into twenty-eight languages. Flexbox Froggy[56] and Grid Garden[57] challenge visitors to learn CSS concepts via open source browser games. These acts of creation directly educate visitors. Stephanie Eckles resurrected Dave Shea's CSS Zen Garden[58] via her StyleStage project,[59] inspiring a new generation of experimentation, collaboration, and modern CSS tinkering in the open. All of these efforts are great applications of the open source model to not only teach via their content but also through their process. They demonstrate how to engage at an impactful level of the spectrum to benefit us all.

Fund

Free and open source software (FOSS) is sometimes anything but. There's plenty of money to be spent and solicited for. Funding for the rank-and-file maintainer means anything from buying them a coffee to signing up for one of the many sponsorship routes available to projects and individuals. This value exchange replaces time for your pocketbook, yet the reciprocity can vary from project to project.

Do your research before sponsoring a project. Can you see an accurate accounting of project income and expenses? Some platforms like Open Collective[60] answer questions like, "how do they distribute funds to their contributors? At what interval? What other expenses are there?" I audited a few larger projects and saw them sitting on a hoard of recurring monthly donations. That doesn't feel like a successful disbursement plan. If you donate, do you get anything in return for your sponsorship, such as priority support, advertising, or kudos? Patreon[61] provides built-in mechanisms for private content and tiers of rewards to funders. GitHub has sponsors-only features too.[62]

During a time of socio-economic upheaval caused by the COVID-19 pandemic, maintainers of the Jest project created public and private bug bounties. This enabled new and existing contributors to tap into Open Collective funds and in a small way supplement their disrupted income streams.[63]

Funding is not always a crowd-sourced endeavor, however. Organizations and companies can invest in projects, providing sustaining contributions and monetary aide for a larger impact. The OpenJS Foundation has twice partnered with the Open Source Security Foundation to sponsor significant investment in foundational ecosystem projects like Node.js and jQuery.[64] Why these? Well, when your library is installed on over sixty percent of the top million websites, you become a project worth protecting.[65] Government entities like the European Commission have supported hackathons and bug bounty programs to buttress open source projects across the globe.[66]

Come As You Are

We can only drive in one lane at once. I don't expect or encourage you to do all of these things at the same, or with every project. It wouldn't be healthy or sustainable. That road leads to fatigue and burn out. There is no explicit drive to *climb* the spectrum as if it were a mountain or linear journey. If that happens as a by-product, great, but it's not the goal. I've outlined these avenues of engagement not to intimidate, but to inspire and re-frame open source contribution as more than hacking with a hoodie on in the dark of night.

All along the spectrum of engagement opportunity to work within open source lies. No type work is of more value than another. Mandy Brown writes in "No one is 'non-technical'"[67]:

> *The categories of "technical" and "non-technical" serve wholly to privilege those in the former, at the expense of the latter. [...] The "non-technical" nomenclature not only does a disservice to that work—and to those people—it also diminishes the ability of the organization to really get the most benefit from those skills. It's difficult to earn value from the things you disparage, even if the disparagement is superficially polite.*

Brown says that product teams, and by my extension, open source projects, would fail without the spectrum of engagement. The skills, temperament, and behaviors essential for a large software product to succeed must be derived from more than just the code. Acknowledging this need for balance, the All Contributors project[68] helps recognize the multivariant output an individual contributes to communities. It offers a GitHub bot and CLI enabling maintainers to celebrate community work along the spectrum of engagement—at the time of writing thirty-three distinct activities.[69] You don't need a t-shirt, conference swag, or a bumper-sticker-laden MacBook to prove you are an open source practitioner.

Instead focus on the value you conjure, incremental and right-sized for the moment. That might be intrinsic, kept close to you alone, because it's 4:14 PM and you have to save your work before getting to the bus stop, starting dinner, and cleaning the house for your in-laws' visit tomorrow. Life is hectic and we should give ourselves grace in juggling it all. Folding the laundry can wait, as can open source. But maybe a spare moment lets you figuratively match and fold the socks: you send critical feedback to a project while you are using it, contribute that documentation fix or issue +1, or even ask a question yourself. You are furthering the discourse of the project with any amount of this work, however mundane, creating value from your time.

Chapter Three

Consuming Open Source Software

The thrill (or terror) of a blank file never quite goes away. You stare at the void, and the void stares back. You, at the start of a project, or a new ticket from the backlog, are met with boundless options to get the job done. Or instead of an infinite canvas, you see a brownfield duct-taped one-of-a-kind invention that is your ecommerce frontend. Looking at it the wrong way might topple it over. Hesitancy creeps into your thoughts, how should I do this? Have we done this before? Maybe you can borrow an existing pattern. Has *anyone* done this before?

The answer to that last question is almost always "yes." It turns out, new problems are rare in this field. Rarer still are new solutions. Can we integrate something already written? You're about to invite a powerful multiplier into your codebase—for better or worse. Its name is dependency.

Dependencies

The right dependency can propel or imperil your project. Using an open source library frees you and your team up to worry about other goals. You build real value if you meet user (and business) needs within the constraints imposed upon you. And let's be honest—there is never enough time. If we can be faster and more efficient in one area, we accelerate the pursuit of other efforts. This can be a critical differentiator of success or failure. You may hit a deadline to ship a feature by offloading expertise and complexity. Or copy-pasting that previous trick. But sometimes a breaking change blows up your test suite and you need to decide whether or not it's worth refactoring your tests or forgoing the upgrade *again*.

We make these determinations all the time, weighing options and outcomes. We engage in a value exchange with the grocery store to safely procure, store, and sell us lettuce as a direct abstraction of gardening. Likewise, we don't often worry about how the chunked HTTP packets travel from server to client and back again. We trust that abstraction, click by click. We live within this omnipresent balance of trust and risk. We offload cognitive burden or material control so we have time to work on other things. Recall in Chapter 1 when we discussed an estimate that the demand-side value of open source software is nine trillion dollars. That's how much money we'd spend to recreate our usage from scratch. We *depend* on tools and services to help us. But they can become so essential to our days that we become *dependent on them*.

Dependencies. A loaded term, it turns out. So that problem you were trying to solve on deadline? Someone has done the work. And they have open sourced it. And so have thirteen other people. Surely one of these solutions will work, right? But how do you know which one?

Packages, packages, packages

I will be framing the majority of our conversations about dependencies around the predominant package ecosystem of the day: npm.[70] Its size (around three million packages) and entrenchment make it a tempting destination, and make it feel unassailable as a source of solutions. Who knows though, I doubt Bower maintainers envisioned its collapse.[71] This guidance can be generalized to any managed dependency environment... be it Go's modules, Rust's crates, Python's packages, Deno's URLs, or you know, Oz bricks (when the next big thing arrives). Dependencies can come in other forms, too, not just downloaded text files we execute in a trusted environment. *Oh, wait, exactly like that.* Avoiding the npm ecosystem and downloading files from a CDN like unpkg, JSDelivr, or cdnjs is still a form of dependency management, one that is both software-related and network-related.

Let's talk codependence and compromise

We need to adopt a nuanced and ever-present vigilance around dependencies. If we are not careful, we create a conflict of interest when considering them. Declaring this bias at the door is essential to meeting project or user goals. We are collectively compromised by the notion that salvation is at the end of the `npm install` output scrawling by on our terminals, every solution a single `import` away. It's no surprise then that this easy access translates to increased page weight. According to the HTTP Archive, the median mobile page weight has *sextupled* in the last ten years, to over two megabytes.[72] Sure, a lot of that is also imagery, but JavaScript accounts for a median twenty-one mobile network requests for four hundred sixty-one kilobytes, thirty-five percent of which appears unused.[73]

Don't get me wrong—there are some truly magical dependencies out there. Not just toys, but things that solve hard problems. Take Browserslist,[74] a tool that integrates into CSS postprocessors, JavaScript transpilers, and code linters.

You specify a target set of browsers and the integrated tools adjust their output accordingly. Their default maps to `> 0.5%, last 2 versions, Firefox ESR, not dead` to cover most evergreen browsers, desktop or mobile, for around ninety percent global coverage.[75] Say you want to exclude that CSS custom properties polyfill in your app bundle? Take the guess-work out of which browsers map to this feature and specify `supports css-variables`. `browserslist` compares that query with a database of browser features per version and results in at least ninety-seven percent global support. Progressive enhancement techniques shouldn't be ignored either, such as `@supports` CSS rules.[76] With a bit more upfront thought, you can enhance a user's experience if their browser supports it—without the overhead of a dependency. If their browser doesn't, you've already designed an accessible baseline. The approach right for each situation is a matter of consultation and compromise. But as a representation of dependencies offloading complexity, sign me up.

Some equate this over-reliance on dependencies to a "JS-Industrial-Complex," as Alex Russel says, complexity for complexity's sake. This leads to lock-in, specialization, evangelism, and a self-referential feedback loop to instrument, educate, and make sense of more and more layers of logic. JavaScript is cast as both the hero and the villain here. One sells us the problem while the other sells us the cure. But dependencies are tools to be wielded, and us as developers play the role required of us in the moment. I challenge you all to think beyond the false dependency dichotomy. Dependencies are a fantastic means to defer expertise and risk management, accelerate progress, and propel projects forward.

Where do you want to spend your time?

Consider the trade-offs you want to offload to someone else's scrutiny by taking on a dependency. You might not have capacity to write test coverage for every edge case your solution encounters. Accounting for every circumstance is expensive and must be weighed against the probability and impact of occurrence. What is the consequence if your software is wrong? Can it be right... ninety-nine percent of the time? Perfection and precision is an admirable goal in most things, but it comes at a cost. The now-deprecated `Moment` timezone library runs over one million unit tests in the browser.[77] They run them so you don't have to. (FIG 3.1)

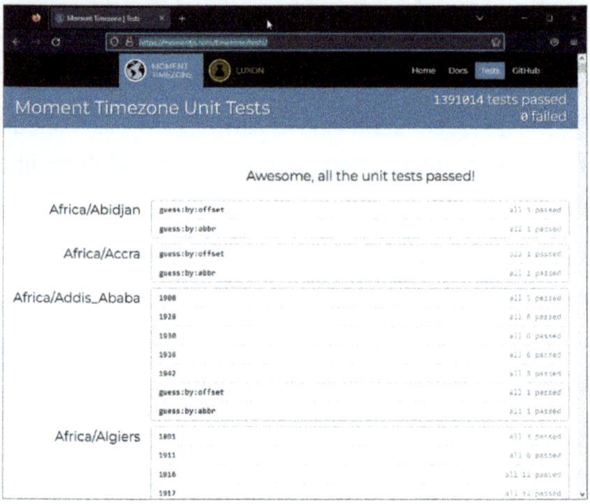

FIG 3.1: In-browser unit tests for the Moment timezone library. All 1,391,014 of them.

This makes sense when you think of the Sisyphean task that is their mission—to keep in synchronization their library with the world's decreed changes in time zones,[78] such as the mess that is Indiana,[79] or when Chile changed their daylight savings time to not disrupt an election. Thankfully, a standards body funnels these changes to downstream tools as a dependency. Libraries, runtimes, and browsers adopted this standard within the `Intl` API,[80] distributed via engines like Chromium or Node.js via the Unicode Common Locale Data Repository,[81] the International Components for Unicode library,[82] and the time zone database maintained by the Internet Assigned Numbers Authority (IANA).[83] Open source and public domain all the way down, with communities dedicated to keeping us all on time for that next Zoom call, PTA meeting, or dentist appointment.

If you write in-house code, does it need unit test coverage to build and maintain confidence? If I had to pick one, confidence is more important than coverage. Entire books and methodologies swirl around the Internet regarding tests, so I won't attempt an exhaustive run down. But I will ask, how much do you need to know the code is working as designed? Try to stay out of the implementation details and focus solely on output. If your code doesn't have output, consider how to rewrite it. Emphasize functional test suites that mimic real-user behavior. Watch out for tests that stray into someone else's code. You don't need to assert that Moment.js is working right. Remember the million tests they run? They got you. And yes, sometimes a missing test or use case is the perfect avenue for engagement with an open source project, but only go there when you've explored options or the omission carries impact and degrades your confidence.

Tech debt

The more code we write, the more code we need to maintain. Dependencies give us an enticing shortcut—and as cuts go, short ones aren't always among the worst. They are an outstanding way to graft disparate work together into what we pass off as a whole entity.

What's important is a clear vision of your dependency as a form of tech debt. You may have heard that term before and drawn negative connotations. I think that is mostly fair as a lasting impression, but the reality is more nuanced than that. Tech debt is an accelerant to progress. Taking a dependency offloads work from already busy and burdened teams, a path to get where you want faster. The value exchange is clear: code now in exchange for reliance and patronage. It's a form of indebtedness. If you've ever purchased a car, a property, or anything really on an installment plan, you are familiar with debt. It is prohibitively expensive to save up the entire lump sum for a purchase—you may be saving your whole life if you needed cash to purchase a home. (Look up the etymology of *mortgage* sometime.) But a line of credit extended to your family brings your goal within reach. We cannot function without debt.

The size of the debt is important to consider, both financially and when taking on a dependency. Small dependencies and APIs are quicker to swap out with a replacement or in-house solution. Large entities like a JavaScript framework might be harder to replace. You may grow accustomed to the amenities and ignore compromises you are making. Vendor lock-in via hosting deals, poor separation of concerns, and freemium adoption are all too common. The salesperson was quick to remind us of the nineteen cup holders on the SUV we briefly had, but downplayed the electrical problems that caused a dead battery six times in a year.

Many dependencies bring not only themselves, but their own baggage—*transitive dependencies*—a hidden clause in the contract that can sneak up on you and ruin a deployment or a day. We reduce the risk of tech debt default by keeping a clean and honest ledger of where our compromises manifest. Pay down your debt with version upgrades, testing, and continual evaluation of the dependency landscape. I suggest investing in tools tools like `size-limit`[84] or Lighthouse CI[85] to help define and monitor performance budgets as your codebase matures. You may need to rely on a dependency indefinitely, and that's okay. Just be mindful of the choices you are making in the moment when that install command is on your clipboard—what value exchange am I getting here? What compromises am I making?

Another kind of compromise: supply chain instability

Open source code dependencies couple your project to the outcomes of others' work. This work isn't your coworker Kyle's bad push that had a merge conflict in it. This work is unseen, four levels deep in our dependency tree, published at all times of the day, and hard to plan for. We expose our solution to upstream events, and if we are not careful, can be negatively affected. Gosh, what a downer this whole chapter has become. By naming our fears, we can begin to understand and overcome them.

How entwined are dependencies these days? You can think of your project as a graph: each module represented as a point with declared dependencies as lines leading to other points. From your initial set of dependencies might sprawl out a veritable nest of lines, localized centers of activity, and even duplication. Maintainers that understand this often endeavor to decrease or remove dependencies from their offering. The less they have, the less they pass on to adopters. Frameworks, by their nature, bundle library functionality with conventions to create their value. But they too benefit from a reduction in dependencies. Let's peek within the React ecosystem and compare Create React App[86] with Next.js.[87] The latest version of Create React App contains one thousand thirty-seven dependencies, (FIG 3.2)[88] compared to twenty-three for Next.js. (FIG 3.3)[89]

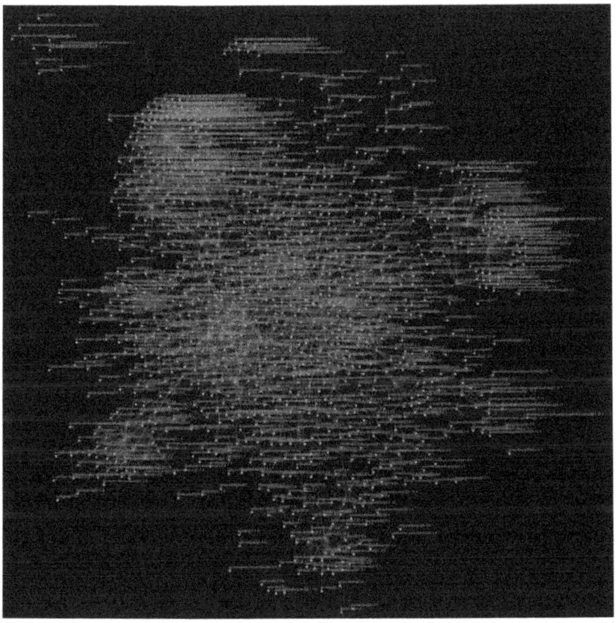

FIG 3.2: Create React App's "react-scripts" dependency tree visualized—one thousand thirty-seven dependencies in total.

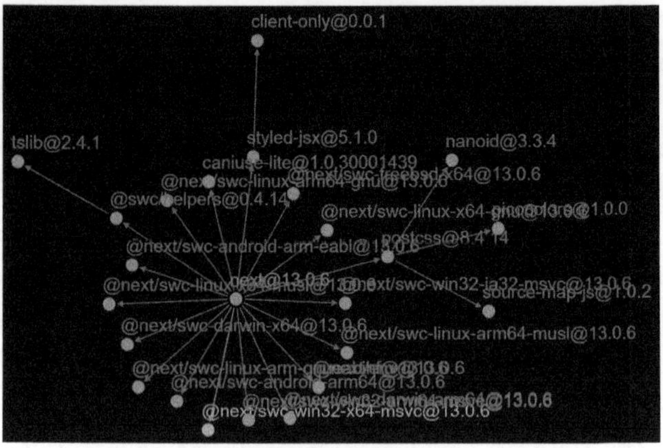

FIG 3.3: The next framework dependency tree visualized—twenty-three dependencies in total.

Earlier versions of Next.js had hundreds more—so a concerted effort was made to internalize and reduce dependencies. Next.js's internalization comes with a compromise, however. The more secure, stable bundling translates to larger binaries. These projects offer similar functionality with vastly different dependency postures. You can imagine what could go wrong with so many pieces of software crammed together like a clowncar.

If you are having trouble imagining, we don't need to look too far. These dependency relationships create wobbly block towers. Sometimes they fall over spectacularly, without warning, and cause a great deal of noise.

- The PostCSS ecosystem deleted a file during a patch release. This brought down dependent tooling, such as Create React App.[90]
- Another accident occurred within `is-promise`, when the maintainer switched bundling conventions without understanding that it would alter `import` / `export` mechanics.[91]
- The Nx monorepo management tool released a minor update that removed its own bundled dependencies, omitting an automated migration to add them back into user systems.[92]

Events like this are more common than I'd like to admit, but then they are over. Usually the problem is discovered quickly, and a fix pushed in a matter of minutes or hours. This is open source Jenga, the tower toppling over. We all agree to pick up the pieces. Despite the annoyance, when things fall over, the community corrects.

More concerning is when developers understand how fragile our dependency foundation is and seek to weaponize or exploit it. The infamous `left-pad` incident[93] affected projects the world over when its maintainer in protest unpublished every npm package they controlled. Missing dependencies caused thousands of cascading failures as dependent projects looked in vain for the library. The incident and aftermath led to improvements by the npm registry to prevent this in the future. It also highlighted again the compromise developers make when taking in dependencies—the `left-pad` algorithm isn't novel or intensive to write (in fact now included in JavaScript as `padStart`), but it's still possible to offload it to someone else. The *Faker* and *colors* libraries were similarly hit with intentional sabotage,[94] their maintainer citing unsustainable funding models. Targeted sabotage emerged as a form of protest after Russia illegally invaded Ukraine too.[95] Worse yet are malicious actors publishing packages intended to ensnare unsuspecting users, either through typo-squatting,[96] account takeover, or dependency confusion.[97]

These risks strain the open source contracts we work within. Luckily, we can take steps as consumers and maintainers to safeguard our code. The dependencies your project uses (or doesn't use) are just as important as your first-party code. Exercising prudence in adding new dependencies will always pay off. Know when to dial up the scrutiny. Be excited that someone else has solved that problem for you, but make clear accounting of the tradeoffs you tally in taking a dependency.

Do I Even Need a Dependency?

I've been talking about dependencies as if their inclusion in a project is a near-forgone conclusion. The open source landscape creates this illusory abundance for us. But know that there is always a choice, that near-hidden value exchange we engage in as consumers. We can choose to *not* take a dependency and therefore retain utmost control. The aged but still relevant You Might Not Need JQuery[98] and You Might Not Need JS[99] sites are testament to the options before us and the trade-offs we make.

We constantly delegate complexity out of our minds or adopt shortcuts to focus on what matters more. This is true for getting coffee as it is choosing a date library. We interact with these complex ideas via neat, well-defined APIs. But there is choice to be had in the APIs we integrate. `driveToStarbucks()` is an opposing implementation to `brewPot()`, though the output of caffeine is the same. Why? The cost and carbon footprint couldn't be much more different—the value exchanges are imbalanced, (FIG 3.4)[100] as are the page weights associated with too much JavaScript.

FIG 3.4: Abandoned mobile-order drinks at a Starbucks after they got reportedly an hour behind fulfillment. With "free" refunds, what's the cost? Photo posted by LI0NHEARTLE0 on Reddit at https://www.reddit.com/r/pics/comments/yxu5e6/

Let's analyze a theoretical dependency decision. Say you need to add a feature to a wedding website, counting down to the big day. As a UX enhancement, we want it to display a human readable amount of time, such as "in 5 months." As the date gets closer, this will change to "in 17 days," "in 1 day," maybe even, "in 2 hours". We have a lot of options at our disposal here.

You might reach to the popular Moment.js library to start. The `moment(weddingDate).fromNow()` function will accomplish this for us as a convenient and expressive one-liner.[101] But the whole library comes for the ride, and if not optimized, takes a whopping 1.4 seconds to download on a slow 3G connection.[102] We can do better. Admittedly, Moment's maintainers agree (the libary is deprecated).[103]

An alternative library exists, Day.js, ninety-six percent smaller, taking fifty-seven milliseconds on the same connection constraints.[104] That's twenty-five times faster. Except for when you want this relative time feature, then you need to load a plugin too.[105] The composability is something you opt into, rather than getting the whole kitchen sink. That might be a good enough balance for most products. Have that conversation with your team—and as Dan Mall says, "decide in the browser."[106] Does it work on the device classes you care most about?

But further, we should wonder, could we write this code ourselves? Could we do it as efficiently as Day.js is doing it, while implementing forty-six assertions of accuracy via automated unit test?[107] Perhaps the answer is yes, since our hero API the `Intl.RelativeTimeFormat`[108] method makes this a native feature available to around ninety-five percent of devices globally.[109]

I wrote up my own version of this feature, working my way through encountered pitfalls as if I were writing it for a client or a product story.[110] It wasn't *hard*, but it was nuanced.

- `RelativeTimeFormat` expects you to provide it the value as an integer and the unit of measure. So I first had to compare dates using epoch time, and then hold onto the difference between dates as milliseconds.

- For the unit of measure, I need to do some quick math to figure out what ballpark of time our timespan is. Are we talking years, months, days? Do we care about hours and minutes? I didn't handle those, but a good implementation would.

- I decided the formatter should look forward in time and also backward in time, so the difference value timespan needed to be revisited to provide a positive or negative value to the API.

- The end result, minified and gzip encoded, measured two hundred ninety bytes.

As a one-off feature, perhaps this is adequate. And it should have test coverage in an ideal world. Recall the question, "where do you want to spend your time?"

Consult your own analytics, use cases, and desired user experience to inform your decision.

Despite the standardization of `RelativeTimeFormat` making its way into browsers, Moment.js is the eleventh-most-used npm dependency according to the Linux Foundation / Harvard Census released March 2022.[111] Its ubiquity might be a lagging indicator of project downloads, but the site npm trends.com doesn't indicate a decline even in the face of competitor growth.[112] Why? Well, it's easy to pick on a single use case like my wedding tracker feature and find a native solution—but can we so quickly match the breadth of Moment.js's API? You might find yourself sweating the details of writing a date library instead of a wedding tracker. Maybe I should open source that, you say to yourself...

How Do I Evaluate a Dependency?

Stars, download counts, and the like only tell a story of consumption derivative of popularity. There is undoubtedly comfort and safety traveling in these herds, and we shouldn't discard them as indicators. Ecosystem size has a large impact on availability of help, tutorials, and examples. I know a casual Robert Frost reader would disagree, but you will find more guidance and solutions along the well-traveled road than taking a chance on a low-visibility dependency.

Look to these alternative considerations if a deeper analysis of a dependency is needed. You might find a gem of a dependency on a dusty shelf, just waiting for a chance. Or you might disqualify a choice that at first seemed like a frontrunner. Identify the real risks to your project mitigated by the dependency. Look for net-win value exchanges.

API

Does the dependency do what you need it to? That feels like a safe place to start. Recall that the API is the public contract between consumer and creator. You want an API that is clear, stable, intuitive, and well-documented. Bonus points for:

- **Types:** this helps with editor code completion. TypeScript[113] and JSDoc[114] excel here.
- **CommonJS and ES Module bundling:** this helps with selective integration into your environment. CommonJS format is often associated with Node.js, while ESM applies to browsers and newer tooling.
- **Versioned documentation:** you'll want an easy way to reference what once was, should you fall behind the latest release.

We need to go further than, "does it work?" however. Ask, "does it cause no harm?" An API's efficacy and behavior should never be called into question. When we choose to rely on someone else's code, the trust-fall exercise extends into our development environments, our CI systems, and more importantly our end users' devices. The dependency should be as easy to remove as it was to add—lest lock-in or compromise creep into the codebase. Michelle Barker documents this struggle with Tailwind on her blog, *CSS { In Real Life}*[115]:

> *when evaluating the use of a dependency that has such a large reach within a project, it's wise to also consider what happens when that dependency is no longer useful to us, or when a new release contains significant breaking changes that it might necessitate a large refactor to keep our project up to date.*

If you think evaluating a dependency is a stressful task, know that it is nothing compared to the dread of feeling stuck within a specific technical path. Make sure your process drives the tools, not the other way around.

Perhaps the most famous API disaster of them all was "smooshgate"—a well-intentioned API decision so bad Google wrote a fifteen hundred word FAQ about it.[116] The open source library MooTools[117] established an array method called `flatten` which reduces an array with nested arrays into a single array. (FIG 3.5)

```
var myArray = [1, 2, 3, [4, 5, [6, 7]], [[[8]]]]
var newArray = myArray.flatten() //newArray is [1,2,3,4,5,6,7,8]
```

FIG 3.5: Demo of what the MooTools flatten method does to a complex array.

This sounds useful enough. So useful, in fact, that it was proposed[118] as an addition to the ECMAScript 2019 specification (the standard defining how JavaScript works).[119] When early previews of the spec landed in browser nightly builds, however, long-stable websites started breaking. Why? For some eight-plus years, Mootools implemented `flatten` (and many other methods) by directly adding to global `Object` prototypes like `Array`. This alters all subsequent instances of the object, leaking a proprietary implementation atop the standard. A long tail of old versions of Mootools were out in the wild, unmaintained, included on websites and CMSs, etc.. It was impossible to substantively address what would have been ecosystem-level instability. End users would suffer, running against the ethos of "don't break the Web" that enable us all to still enjoy the first website,[120] or the Space Jam homepage from 1996.[121]

But why the *smoosh* in *smooshgate*?

A contributor to the spec named Michael Ficarra proposed the rename from `flatten` to `smoosh`[122] as a light-hearted solution. In the end, the proposal landed on `flat` instead of `flatten` (or `smoosh`). When your API choices impact the very evolution of the programming language, you've reached record-levels of engagement and value exchange (but not in the good way). It's fun to jest in hindsight, but we still aren't past these basic concerns. The React library is "instrumenting" the global `fetch` function under certain circumstances.[123] What could go wrong?

Michelle Barker gives us this final gift about API scrutiny, from the same post as earlier:

> *By comparison, web standards evolve relatively slowly. But they evolve slowly for a reason: features that are added now need to be supported forever. They are not designed to become obsolete.*

License

The open source license a project adopts cannot be understated. Permissive or restrictive licensing models prohibit or secure certain rights when distributing software. They are value exchanges that went to law school. You don't want to be near or past the finish line on a release only to realize you are in noncompliance with a license. Determining the license of a dependency is usually straightforward, but there are a couple different places to check:

- A LICENSE file at the root of the repository
- A `license` key within the `package.json` file
- The README

The lack of a standard can confuse tooling designed to determine compliance, so you need to be thorough in your search. For example, `npm view` will only return values defined within `package.json`. A tool called `license-checker`[124] will crawl your project's `node_modules` directory and use a heuristic process (guessing) to find the license of each dependency. This can be useful for a broader review of the code comprising your work, not just inspecting a single dependency. (FIG 3.6) You can filter certain licenses, direct output to different formats like `json` or `csv`, or fail the command outright on the discovery of specified licenses.

```
eleventy
⚡ npx license-checker --summary
├─ MIT: 410
├─ ISC: 66
├─ BSD-2-Clause: 15
├─ BSD-3-Clause: 10
├─ Apache-2.0: 7
├─ (MIT OR CC0-1.0): 3
├─ Python-2.0: 1
├─ CC-BY-4.0: 1
├─ MIT*: 1
└─ 0BSD: 1
```

FIG 3.6: license-checker output for the Eleventy project. The tool deeply inspects all dependencies and reports the license groupings and counts.

Some licenses like the GPL[125] require you to disclose the source code of any distributed work relying on it. Being cognizant of the licenses laid as the bedrock of your project's dependencies is an important determinant of its own capabilities, liabilities, and perhaps marketability. Who maintains and governs the license is important too. The gold standard is a project governed by an open source foundation. Be more wary of the single company or single individual projects—their motivations might shift out from under you. You don't want licensing to be an afterthought with an angry engineering manager, legal counsel, or client. We'll cover this topic more fully in Chapter 4 too.

Longevity

The age of a dependency says a lot about its resilience or adaptability to changing trends in the marketplace. A flash-in-the-pan technique or now-fallen-out-of-favor approach both present as fleeting sizzle. But a project like the Express web server,[126] for instance, has a slow-burn staying power that is evident in its fifteen-plus year history. Competitors arrive to usurp Express, even from its own creators, but none have dented Express's pervasiveness. Which is weird when you look at their analytics—it seems like the pace of development has slowed to a crawl. I've had the thought: *is Express dead*? Turns out that when an API is mature enough it doesn't need more enhancement—there just isn't much more room to grow without changing the intent of the module. In some ways, Express isn't *dead* as much as it is *done*, a testament to its value and relevance. Look for dependencies that have seen and survived a few hype cycles.

Stability of release

When evaluating a dependency, check its CHANGELOG or releases page. A steady, high-frequency of release is almost always a hallmark of a well-run project. Automatic merge and release strategies will generate a lot of activity, sure, but be on the lookout for something rotten with all the releases: regressions, botched artifacts, and costly cleanup.

Careful and considerate maintainers will take steps to ensure the stability of their toolset, shielding the community from unexpected change. This is the essence of *semver-compliance*, which we'll talk more about in a bit here. Look for patches that contradict previous releases, or issues that mention a breaking change on a non-major release. I've mentioned now a few times "the speed of open source" and again here it's like a skateboard on a hill. There is fun and fast and then there is out-of-control danger.

Community safety

Download counts and vanity metrics won't tell you the complete story whether a project is safe to use, either in the code or at the watercooler. Your favorite library may have a vocal minority of toxic fanboys eager to take one person's preference as an affront to theirs. Remind them: it's okay to like different things! Ask around. Check hashtags on social media, industry blog comment sections, and issue logs. See what folks gravitate to the community created by the dependency. Actions speak louder than words here. Does the project have an enforceable code of conduct? Are maintainers role models in setting the mood? Does it have inclusive representation worthy of your time? Can you envision engaging without flinching at the thought of responses? We hope to find projects that can meet our measure. And once we do, we need to be mindful consumers.

Responsible Consumption Habits

The dependencies we take onto a project must be managed sustainably. They are like any other tool we must acquire and organize, be trained on and teach, secure and extend to our will. They cannot rust from disuse, nor crowd out the workspace. The wrong tool applied liberally can complicate timelines and endanger the final product. You'll know this if you've ever tried killing a fly with a hammer. We now have a more nuanced view of when we need a dependency and how to evaluate which one is right for us. How can we ensure our carefully considered open source dependencies remain under control?

Semantic versioning

One of the principal ways we can keep dependencies in check is to thoroughly and rigidly follow semantic versioning, or *semver.*[127] I've used the term a couple times already, so it's worth level-setting now.

Semantic versioning within your project will most likely manifest within Git tags or the `package.json` file. Dependencies of your project will be defined in the latter, with any number of named modules accompanied by a version. The version can follow a couple different conventions, but most look like this:

MAJOR.MINOR.PATCH
1.3.4

This semver is a promise that the `1.X` line of releases will always have a compatible API with one another. It's a value exchange made numeric. If the promise is upheld by maintainers, upgrade the MINOR or PATCH version as your needs dictate—but exercise caution when upgrading the MAJOR. Look for changelog entries, release notes, and migration guidelines that help you navigate the changeset.

Semver helps you understand how change is introduced into your codebase. Whenever I field support questions, the first thing I establish either from questioning or commit analysis is, "what's changed recently?" Many times an update to one or more dependencies is a culprit. Dependencies that do not follow semver should be viewed with scrutiny, suspicion, and reluctance. Unfortunately, you may not know the stove is hot until you get burned once. Then, you start to look for the signs of a healthy project. Incrementing semver releases. No hedging or wordsmithing about breaking changes announced in smaller versions. No complaints in an issue log of unintended behavior on upgrade. Semver is just a number—maintainers shouldn't hesitate to communicate via its changing. The consistency and adherence to these concepts is as important as traffic lights are to the safe passage of vehicles. If one person breaks the rules, they may cause harm to themselves or others.

Our tooling assumes that maintainers follow semver. The `package.json` file describes the tools, dependencies, and configuration of most modern front end projects. It's the recipe or owner's manual you turn to when orienting to a new environment. We'll use this `package.json` file as an example. (FIG 3.7)

```json
{
  "name": "myLibrary",
  "scripts": {
    "start": "node fetchList.mjs",
    "test": "jest",
    "lint": "eslint ."
  },
  "dependencies": {
    "axios": "4.0.0"
  },
  "devDependencies": {
    "eslint": "^10.0.0",
    "jest": "82.7.86"
  }
}
```

FIG 3.7: A fictitious package.json with a precariously careted devDependency on eslint. A lockfile is the only thing keeping that declaration deterministic.

You might recognize our `fetchList` file from before. The code requires the Axios library to run, so we define that within the `dependencies` key (all at currently fictional values). The testing library we used, Jest, is still there too, but defined within `devDependencies` because it's only used for local development and not distributed to consumers if `myLibrary` is published. Notice the code quality tool ESLint as the most notable addition. To get all these tools to run locally, you typically run `npm install` (and variants you can explore).

What happens during `npm install`?

When `npm install`[128] runs, the contents of the project's `package.json` are read. The algorithm looks for dependencies to download and install. Every dependency is checked for dependencies itself, recursively, until a tree is formed of all project requirements. If a lockfile is present in your project, it is used to resolve this tree faster. Think of a lockfile as a pre-cached tree of dependencies. Sharing the lockfile with your team via Git makes everyone and your build server use the same deterministic set of dependencies too. Either way, the resolved dependencies are downloaded and installed in the `node_modules` directory, their optional bin scripts made available to the project. The lockfile is updated if dependency definitions or versions were altered in some way. You can configure a project to error during install if the lockfile and `package.json` dependencies get out of sync. Variant package managers like Yarn or pnpm have their own flavors of this algorithm. pnpm in particular goes steps further than npm in isolating modules from one another, and creating a local cache of downloads. Once downloaded and installed, dependencies are available for use within your project.

Semver-compliant algorithms

Why the crash course on `npm install`? Well, because it gets more complicated. Built into semver and the npm installation algorithm are some conventions enabling developers to define *version ranges*. The two we highlight are the tilde(~) and the caret(^). The tilde matches MAJOR and MINOR versions, while the caret matches MAJOR versions. *Absent a lockfile*, these ranges allow npm to inspect each dependency's published versions and pick the latest release that satisfies the range. For example, if `eslint` is defined within the `package.json devDependencies` section as "`eslint": "^10.0.0`", we are stating that any version of `eslint` greater than or equal to `10.0.0` but less than `11.0.0` is acceptable. You might want to do this if you trust the semver compliance of a dependent project. You are opting in to automatic bug fixes and non-breaking enhancements. Powerful. Useful. Dangerous.

Nevermind all that, pin your dependencies

Why dangerous? This semver stuff doesn't seem like a MAJOR deal. But, I suggest, that any project of sufficient size, team, and complexity will run into dependency management issues. It is inevitable, despite our best intentions. What could happen?

- Merge conflicts
- Local installation hiccups
- Major dependency upgrades
- Surgical lockfile-only patching of transient dependencies via security tooling
- Inconsistent install methods
- Switching branches

Each one of these mundane events increases the likelihood that a developer will want to blow away their lockfile. They will do it, because it's an easy troubleshooting step and feels safe to do when panic strikes. The text scrolls by fast and they feel in control. But when they do that and re-run `npm install`, the resolution and installation algorithms run again with no project-level context. Suddenly the `"eslint": "^10.0.0"` version that was resolved and locked to `10.1.1` a moment ago has upgraded you to `10.6.0` complete with new lint rules that break your CI system. It's a subtle and simple way to bork up your day.

The determinism of a lockfile can extend further into the dependency tree, into transient dependencies too, so there is value in checking a lockfile into Git and attempting to keep it in sync. Yet I've even seen teams `git ignore` their lockfiles. We should go a step further and always "pin" our dependencies to explicit versions within the `package.json`. Pinning removes the caret or tilde characters that are establishing a semver range. If your project has an `.npmrc` config file[129] at the root, adding `save-exact=true` will automatically make pinned dependencies anytime a new dependency is installed. It's a measure better, and an important one. Pinning isn't limited to npm-based projects either. For example, Rafael Gonzaga shows why GitHub Actions should be pinned by commit hash.[130]

Beware the zero dot release

Exhaust other options before adopting a `0.X.Y` release of any software—this is not stable and could change at any time. The default npm project initialization used to be `0.1.0`—which always felt like a curious choice because it suggested uncertainty within API intent at the outset of a project. While that is likely true, I feel like it's led to some high-profile examples of successful projects amassing huge adoption while not communicating properly with semver when they reach a threshold of functionality most would consider "general availability." Simply put: people are using unstable code and no one minds.

A notable example was axios[131] (until it released its 1.0.0 version in October of 2022), which incremented for years all the way to `0.28.1`, each new release a potential quagmire to navigate. The fact that this library attracted tens of millions of weekly downloads with this release posture might be the largest collective fingers-crossed moment in the JavaScript ecosystem. Zero dot maintainers are either intentionally communicating risk (ok), are ill-informed of the trade-offs (sigh), or are willful in the lopsided value exchange they are setting up (yikes). Open source is like that though; some projects will go their own way, and we can choose to engage or vote with our feet. Even when traffic lights contradict the norms, the traffic flows nonetheless.

Dependency management tools

With such an overabundance of npm dependencies at our disposal, the ecosystem has built tools to cope, lest it collapse in on itself. Luckily, automation can help us keep up. GitHub's Dependabot, for example, will tell you when a dependency is out of date or vulnerable, and even submit a patch for you. With any tool, however, misuse and circumstance can combine for negative consequences. Many security vulnerabilities reported in dependencies are unexploitable in practice and impossible to remediate, a potent combination of boy-cried-wolf seriousness. This always-alert posture is exhausting and unsustainable—fatigue will make you complacent over time. I suggest tuning any automated tool to reduce noise:

- Inform and act upon HIGH or CRITICAL vulnerabilities. Handle lower severities during an established cadence of maintenance.
- Batch and schedule automated updates so you can plan for their review.
- Invest in more robust testing coverage to ensure automated updates are safe to integrate.
- Be wary of updates that only impact a lockfile—they will regress with any lockfile deletion.

There's still plenty to celebrate in this space. That such deep insight exists into these complex ecosystems is exciting. Many tools have broad language support now too, offering scopes beyond npm into Docker images, Nod e.js versions, Ruby, Java, Go, etc. See the Resources at the end of the book for some specific tools I've found useful.

No matter the tool, periodic dependency updates you can set your watch to position your project for the most success. Monthly or quarterly updates seem to occupy a sweet-spot balance regarding a cadence. Wait too long and public documentation may drift. It's distracting and noisy to update all dependencies all the time, so use your discretion regarding severity, effort, risk, and impact. Critical bugs should be patched faster, but in general, a slightly longer timeline can help smooth out unstable release thrashing. Allowing dependency debt to spiral out of control is detrimental to project security and individual productivity.

It Depends

There is an app for everything, and increasingly, an open source offering to compete with it. Whether you turn to a standalone program or the codebase you compose together, open source consumption affects you. It is often said that software is eating the world. Almost as many folks are now saying that open source is eating the industry even faster. André "Staltz" Medeiros coined the metric "time to open source alternative" and observed that it appears to be decreasing. (FIG 3.8)[132]

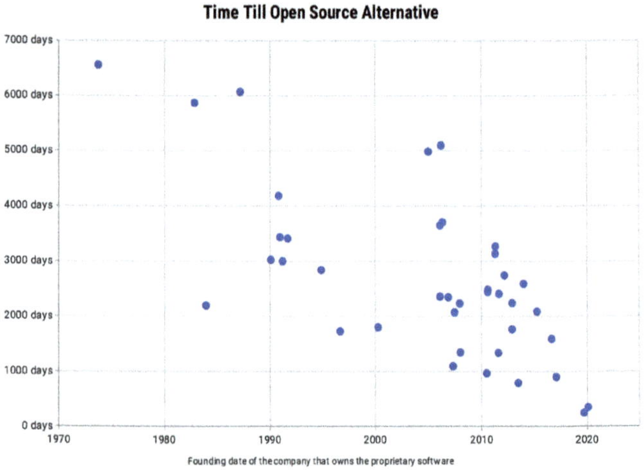

FIG 3.8: A graph showing a decreasing "time till open source alternative" for proprietary software. Image copyright Andre "Staltz" Medeiros is licensed under CC BY 4.0.

The proprietary monopoly of yesterday is gone. The explosion of choice overwhelms the mind and can bring about analysis paralysis if you are not careful. Our landscape is every day more open, more inclusive, and more navigable, while also becoming more complicated, fragmented, and perhaps insecure due to its size and complexity. Can you trust anyone or anything far enough to come to rely on them? Will it be around tomorrow? Next week? Next year?

And yet it's clear from nearly every professional and personal experience: we depend on one another. The family needs the grocer who needs the co-op gardeners who need the farmer who needs the machinist. It's a lie to think otherwise. Projects depend on one another. Systems depend on one another. Economies depend on one another. We cannot escape interconnectivity.

This interconnectivity is a gift and a burden. We are delighted in the shortcut, the serendipity, and the solution solved faster. The vulnerability we open ourselves up to when engaging and depending on relationships is what makes it special. And it is risky. If we are fortunate and discerning, we can choose the degrees of quality in these relationships and abstractions. Which yields better results? I said before this is a false dichotomy, a strawman argument. Don't fall for it. Instead, seek the value exchange beneficial to the moment. Seek the dialog about what's right for the project now, and into the future. It's tempting to repeat the programming joke, "it depends." And I think you'd be right in the declarative sense.

So let's embrace dependency when we need to, with careful craftsmanship. Let's with intention engage. If and when the time is right, you'll find that thing you want to open source. Who knows, one day we may depend on you.

Chapter Four

The Four Files of Any Open Source Project

No matter the size of an open source project, there are norms that newcomers and seasoned developers alike will expect to find within your repository. Think of the products you encounter outside of tech—food, clothing, or hardware. We expect a tray of Oreos to suggest a serving size (so we can ignore it). We read the labels on garments so we know what size to buy, and how to care for them. Installation instructions tell us what tools we need, how to get started, and what to do if things go wrong. (FIG 4.1)

FIG 4.1: Ikea's instructions provide requirements, warnings, and guidance without any written words.

Confusing or absent documentation harms consumer agency and safety. Many companies make product documentation accessible via multiple languages. It's hard to know where to begin, so I've defined what is essential to a project of any size. Let's start with four files that act as a foundation for you and potential users or contributors: a `README`, a `LICENSE`, a `CHANGELOG`, and a `CODE_OF_CONDUCT`.

README

The README document is the window into your project. It must juggle multiple competing needs and values simultaneously in order to be effective. And no, we cannot add an auto-rotating carousel to the top. Consider that different personas will be scanning this document in search of information. A developer evaluating your project needs to know what it does and how. A current user may need to dive deep into an API method's return statement. A first-time contributor might look here to find out how to contribute to the project. Programs might scan the file for metadata. Structuring your `README` with intention—maybe even designing it with the spectrum of engagement in mind—can have an outsized impact on your project goals.

What makes a good README?

One of the first recorded instances of a `README` file was within the DEC PDP-10 user group documentation for electrical circuit simulation programs, SPICE, SINC, and SLIC.[133] What they exactly do is a bit beyond my skillset, but in 1974 the guidance contained such nuggets as "many severe bugs. With this distribution I will no longer make any attempt to distribute that compiler nor answer any questions on it." It ends with "Good luck!" which is a decidedly homebrew vibe. To its credit, it contains requirements, constraints, and performance metrics.

Luck is when preparation meets opportunity. Here's my go-to list of `README` contents.

Demo

A screenshot or animated GIF providing a glance at the functionality one can expect. Macs can create a nice screenshot with `command` + `control` + `shift` + `4`, and then hitting `space` to capture the whole window. Code or terminal snippets can be made with a free tool like Carbon.[134] Many of the code snippets in this text were created using Carbon.

Installation

Instructions how to install the project. This can be a multipart or single command depending on your needs. npm-based projects should specify the version of Node.js and npm to use, as these can have impact on how dependencies install.

1. The `engines` field within `package.json` can define your runtime requirements. All modern package managers parse it. npm and pnpm will emit warnings during an install command if your installed version does not match requirements, whereas Yarn will outright fail.

2. Adding an `.nvmrc` file can signal to users what version of Node.js to run. Tools like `nvm`[135] can read this file and wire up the proper versions. GitHub Action `setup-node`[136] will read this file or the `engines` field, reducing the number of magic strings in your code.

3. If your project is a command-line tool that can be installed and invoked in a single action, add `npx`[137] instructions. This downloads the executable if not already present and then runs the command.

4. If your project is large or nuanced, let the user know when install is complete by providing a health check command, or guidance to determine everything is working as expected.

Usage

A user-facing tour of project functionality. Usage implies that the tool is already installed and ready to be run. What does success look like? How about failure?

1. **Invocation:** how do you run the software? Start with a "hello world" or default example.

2. **API:** document the public-facing contracts that comprise your API. Detail the input and outputs. Provide examples. A documentation-first development model will reveal what actions a user can take.

3. **Configuration:** how can default behavior be deviated from? Documenting available options can help a user fit your software to their

use case. Can configuration be shared, inherited, or overridden? Are configuration options validated? Do you have defaults, types, and lists of options defined?

Contributing

Outline how you'd like people to help with the project. We'll talk about tooling that can help here in a bit, so focus more on the process. GitHub, GitLab, and others also encourage a discrete `CONTRIBUTING.md` document, which gives you are bit more focused area to discuss topics such as:

1. Branching strategy and issue etiquette. Large projects may suggest you open an issue first to confirm the behavior, the plan, and claim your interest in fixing it.

2. Development setup beyond user-installation

3. How to run tests locally

4. How production releases are created and published

The Day.js project embodies many of these concepts within their `README`. "Above the fold" on their `README` are links to translations, browser support, the size of the library, and an example of their API. (FIG 4.2)[138] Links to their official website follow, but staying in the file shows installation and API references too. It's a thoughtful presentation of a lot of content. With any project there are competing needs colliding, so pave the cowpaths[139] and don't overwhelm. README documents should have clear, scannable headings to help navigate.

FIG 4.2: The Day.js README represents a well-thought-out introduction to their project, directing traffic for various incoming personas. Perhaps the logo is too big (j/k).

If you are looking for inspiration, browse around. You'll start to notice nuance among projects of different intents and sizes. Katherine Oelsner built an online markdown editor[140] to help create **README** files. The company divio publishes a documentation paradigm designed around four user-centered needs: tutorials, how-to guides, technical reference, and explanation.[141] Ask someone unfamiliar with your project to navigate your **README** for particular tasks. Fresh eyes will reveal opportunities.

Hello? Do you **README**?

Set the tone you expect others to follow with your `README`. Be welcoming and inclusive to newcomers. This is the first impression of your project. Use it to shape your code and community for success. The `alex`[142] tool validates the use of language in your project and calls attention to potentially insensitive word-choice and phrasing. (FIG 4.3)

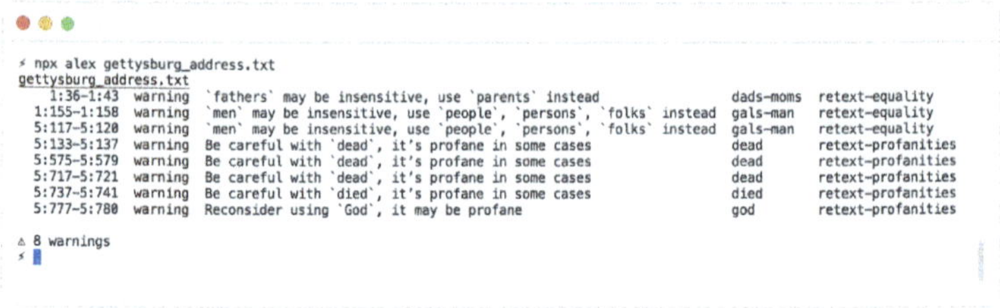

FIG 4.3: alex flags potentially insensitive and inappropriate language.

Many maintainers of projects advertise metadata about a project via shields[143]—dynamic images that display data such as download counts, versions, build status, code coverage, and so much more. The Day.js project earlier used them. Shields give an otherwise vanilla-looking document some visual flair and color, but I caution their overuse. They are far too easy to grow into a bumper-sticker-laden mess.

More importantly, they are usually inaccessible to screen readers like Apple VoiceOver, Microsoft Narrator, and JAWS, as their content is generated on the server with no means to affect the image markup `alt` tag. So we are left with literal read-outs of data URLs (FIG 4.4)[144]:

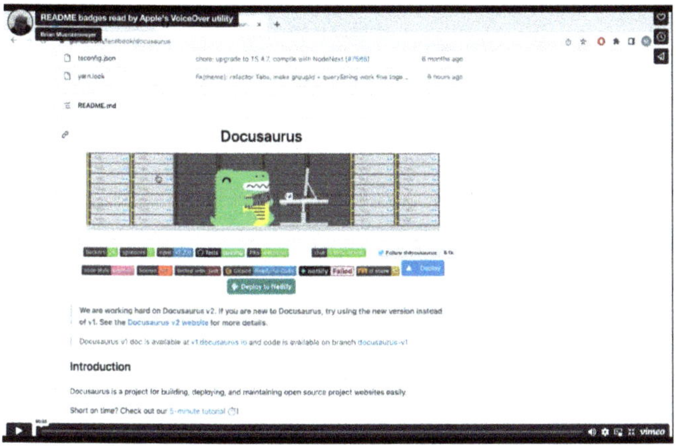

FIG 4.4: A screen recording of Apple's VoiceOver reading inaccessible README badges. Autogenerated data URLs are read verbatim. "Unlabeled image, unlabeled image. Slash 68747470733f2f70...

I cannot believe I have to say this, but... beware vendor lock-in with your **README**. It's outstanding that sites like GitHub render markdown-based **README**s for you and your users. They continue to enhance what you can put into your structured content. There's even a name for it: *GitHub-Flavored Markdown.*[145] Whether they are improving the medium or repeating the risk of browser-specific features is up for interpretation, however.

Take a look at the mermaid diagraming project,[146] which GitHub announced site-wide **README** rendering integration in February 2022. (FIG 4.5)[147] Without extensions installed, popular editors fallback to the text content. This is fine as a graceful degradation, but visiting the same project over on npmjs.com[148] was broken for months. (FIG 4.6) Users encountering your project from that landing-page would be left with a broken or diminished experience you may not have accounted for.

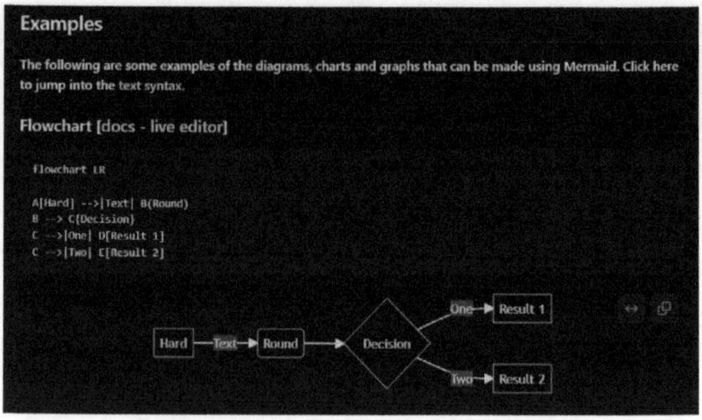

FIG 4.5: A README containing a mermaid diagram rendering on GitHub...

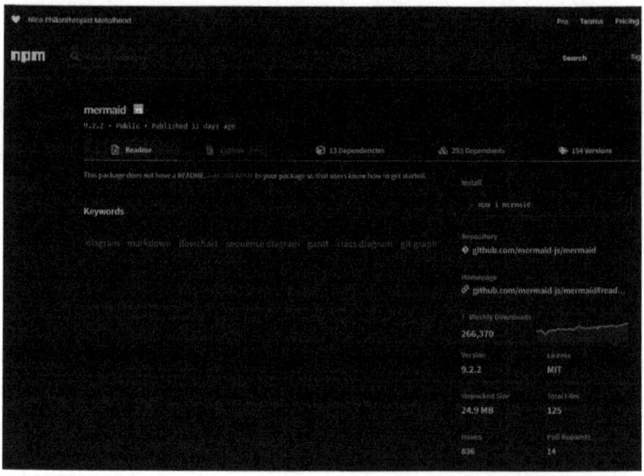

FIG 4.6: ...and a broken view of npmjs.com attempting to render the same content. In an error state, the site logic claimed there was no README at all.

A cherry-picked example, now fixed, you say? This fits a pattern. The site also supports admonition rendering such as "Note" or "Warning" blockquotes.[149] Feedback brought up concerns of accessibility, English-only features on a site with international audiences, and further specification fragmentation. There was also a brief technical preview of GitHub Blocks[150] in 2022 and 2023, declaring that "`README`s don't have to be static." With intentions to further render live demos, project stats, search, sortable tables, charts, and more, GitHub was positioning the `README` as a real homepage. It demoed well, but would other sites be able to render this content, or would editors with markdown previewers? Perhaps they could have pulled it off as a progressive enhancement.

This all might not matter to you, and I respect the net-positive impact GitHub has, but be aware of the value exchange you are creating here—an enhanced experience in one location at the expense of others. Know that this is a convenience not a requirement. When playing in someone else's sandbox, you may someday come to find it hard to move your sandcastle. Jenn Schiffer puts it nicely when she said in a now-protected social media post, "keep an eye out for gardens growing walls."

`README` vs. website

With so much functionality begging to be added to a `README`, it's not a surprise that some projects opt to make a standalone website instead. The sky is the limit in terms of what functionality you wish to build into it. Often these sites become projects in and of themselves to host, maintain, secure, design, upgrade, translate, etc.

Given the landscape, and the ever-present temptation to make a site and devote untold time toward it, I feel responsible to provide some smart defaults in the Resources section at the end of the book. The `npm publish` process will package up a `README` file by default, meaning it is distributed to your end users' machines. The same cannot be said of a website, which requires at least one initial network connection (crossed fingers they make it an offline-friendly progressive web app). I've had productivity stall on a dodgy bus Wi-Fi connection, or roaming while on a phone. So don't discount your offline, lo-tech options when documenting.

LICENSE

Distributing a `LICENSE` file tells other folks what they can and cannot do with your project. The legal side of open source licensing can get complicated really fast. Most people don't think about it until something goes wrong. Let's first focus on the basics here, ease into the complexity, and leave you with resources to turn to when you need to dive deeper. I promise, it's anything but dry.

Projects you put on the World Wide Web retain proprietary copyright until you say otherwise.[151] If you accept contributions in this precarious format, you and the contributor intertwine restrictive copyright and lock-each other out of the car with the keys inside. This may feel confusing as a default, given how easy it is to share code on social platforms these days (GitHub's terms of service don't grant a license to code you host there, only that others can read it).[152] It is best to be clearer in your intent at the outset.

The excellent resource choosealicense.com offers comprehensive education on open source licensing, with an interactive guide to suggest a license based on your values and project goals. (FIG 4.7) I like the user-centric viewpoint choosealicense takes here. The site suggests popular licenses based on the value exchanges important to a project. It also maps existing open source projects to these licenses so you can draw your own conclusions about which communities choose what license. I myself have gotten confused with terms like "permissive" "restrictive," or "copyleft," so let's discuss.

FIG 4.7: The choosealicense.com homepage provides user-centric guidance on which license is right for your project.

Permissive vs. restrictive licensing

Licensing can be described across different dimensions, one of which is permissive vs. restrictive. Sometimes this can feel ill-defined or counterintuitive, depending on what perspective or concern you prioritize. *Does permissive mean people can use it for free? Does it mean companies can sell work based on it? Or is that restrictive?* We can make this clearer. On the permissive end we have the perception of public domain (which is not legally recognized everywhere). Permissive work is free to be remixed, re-sold, and incorporated into other work. On the restrictive end you have individual, exclusive copyright by default. Everything you make is yours until you say otherwise. The fun comes in the middle, where most licenses land.

Licenses:

- **Grant permissions:** what you *can* do with the software. Think "may statements" like, "you may sell derivative works."
- **Define conditions:** what you *need* to do with the software. Think "must statements" like, "you must disclose the source code."
- **Set limitations:** what you *cannot* do with the software. Think "cannot sue" sentiments. This is almost always in the form of an "AS IS" limitation of liability to project authors.

Using these statements as a rubric, we can compare the MIT and GPL license, both prominently displayed within choosealicense. (FIG 4.8)

MIT vs GPL Licensing

License	Can sell and use commercially	Must disclose source code	Must distribute with same license
MIT	Yes	No	No
GPL	Yes	Yes	Yes

FIG 4.8: A comparison of the major differences between the MIT and GPL licenses.

With the MIT License, downstream users can do nearly anything with the code, even close the source and sell it commercially. Conversely, adopting the GPL as any part of your dependency tree is like adding `!important` to a CSS rule—it overrides the cascade and forces all downstream works to adhere to the GPL. This principle is sometimes referred to as *copyleft*[153] (the antithesis of *copyright*). Such-licensed software can still be used for commercial works, but the code must be available and extensible without cost. You can start to see why GPL-centric companies tend to offer services adjacent to the code like consulting and hosting as part of their business model.

Website content or project documentation may fall under the same license as the code, but if you want your technical writing, blog posts, or announcements to have broader reach consider something like Creative Commons licensing.[154] Their license terms and considerations are better suited for non-software works. Vice versa, code isn't meant to be released under Creative Commons licensing, as they don't offer the same permissions, conditions, and limitations that software requires, especially pertaining to distribution and patents.[155]

Phew. I hope that helped! But what if the current license is no longer meeting the needs of maintainers, the community, or defining the right value exchange? Can it be changed? The technical answer is "yes," but let's explore.

Changing a license

I've focused on license differences between MIT and GPL as a reductive primer on the subject, but there are many more licenses. There are so many licenses that we teeter toward duplicity, fragmentation, and unenforceabil ity.[156] Changing a license seems plausible, given the number out there. But once a project is open sourced under one license, it becomes harder and harder to change the license it is released under. Changing from MIT to GPL, for instance, would require the approval of every existing contributor. In general, each contributor to your project is granting their lines of code as a continuance of the current license. You might see this referred to as inbound=outbound. Some large projects try to avoid this consent problem by setting up Contributor License Agreements, or CLAs.[157] A CLA signed by contributors can make future license reassignment easier, among other things. Still following? These are not theoretical concerns. It can get ugly.

In 2018, the monorepo management tool Lerna[158] switched from the MIT license to an exclusionary, custom variant[159] that singled out companies allegedly doing business with United States Immigration and Customs Enforcement, or ICE. This move was in protest to the separation of adult and minor migrants at the US border[160] begun under the Trump administration and continuing into the Biden administration.[161]

Within hours of the change, users pointed out within the issue log that Lerna was no longer MIT License-compliant and should not describe itself that way .[162] Others pointed out that the license change could not be made unilaterally, even by maintainers, without explicit consent from all contributors. There was no CLA on file.[163] One contributor publicly requested that their prior contributions be removed from the code.[164] Amidst the chaos, the original pull request author limited all new issue creation within GitHub to maintainers.[165] They locked, closed, and shutdown attempts to criticize or clarify the licensing decision.[166] Discourse on GitHub, across multiple repositories, and on social media was heated, meandering, and devolved to personal attacks.

The Lerna maintainers and community were not ready to manage the mechanics, discussion, and impact of the licensing change. The move attracted a lot of attention from individuals outside the project on different sides of the political fault lines of open source software. Instigators pushing their own narratives muddied many earnest attempts at discussion. You can find hundreds of comments within contemporaneous issues that I won't include here. Calls intensified to revert the license change and remove the original author of the pull request, citing harassment as a code of conduct violation.[167] Other maintainers did just that: revert the license *and* remove the author of the new license as a combined act.[168] This linking of the legal merits alongside the behavioral concerns drew a new round of criticisms.[169] The code of conduct was eventually updated to clarify responsibilities and outline good and bad behavior,[170] itself a move some claimed to stifle the freedom to debate. In all, the relicensing effort failed to achieve the goals its maintainers intended. Then again, I'm telling you about it, years later.

Events like this fade from memory only to renew, because open source licensing parallels the pace of change around us. For example, licenses may change for less noble reasons than moral resolve, such as a calculated pivot in business-model. Even acquisition. Recent events like Hashicorp's BUSL,[171] Sentry's FSL,[172] or

Redis's SSPL[173] relicensing come to mind. These companies are responding to competitors "free-riding,"[174] setting up SaaS copies of their open source projects. So they relicense, saying anyone but a competitor can use this. Oh, and they get to define who a competitor is. Certainly murky, uncharted legal territory with imbalanced value exchanges. A lawyer may see the worst in that language because the impact outweighs the intent and they are paid to worry. The effect on these projects' communities and consumers is not yet resolved. The decisions may even be justified from a business perspective. Companies must make payroll—and answer to investors—but it's a nuclear option with poor optics and questionable outcomes. Either seemingly forsake your values or reveal how flimsy they were to begin with. The press spins a narrative of open source failure. That the model and movement are incompatible with today's and tomorrow's world. The symmathesy wilts. But I suggest we're focusing on the wrong thing. Tom Calloway sums it up nicely on LinkedIn[175]:

> *i've read a lot lately about how open source is dying, at a crossroads, can't keep up, must be less open.*
> *nah.*
> *open source is people.*
> *people are still capable of amazing things. sure, open source faces challenges, but it always has. as long as there are people willing to collaborate, transparently, in good faith, that light won't go out.*

Open source is people. It always has been and always will be. Whatever the motivation or fallout, these relicensing events highlight foundations' role as uncompromised stewards of open source. They, the foundations, loose federations of people, exist to make projects resilient to decision-making from singular entities. For instance, the Linux Foundation announced support for projects OpenTofu[176] and Valkey,[177] respectively open source forks of Hashicorp's Terraform and Redis's eponymously named product. These organizations respond to instability with a push for open governance, the learning system re-inventing itself in real-time.

In relicensing and fragmentation, projects are potentially alienating the very people that positioned them for success. You can see why this happens and also why communities may be upset. It can feel like a betrayal. The foundational contract of open source feels strained, even broken. We can directly observe big picture societal and market forces colliding with open source ideals. The repo is the battleground; the license in the weapon.

Open source and politics

Author's Note: We're in the license section of this chapter right? Well, my former editors might have wanted me to remain on topic, but this one's too important to omit. Besides, I can break the word count budget when I'm the one footing the bill!
I wrote and rewrote and rewrote and rewrote these next sections. It created the most stress, editor comments, and self-reflection in the whole book. Please know how important it was for me to find my voice here. You can judge me for these words, but I hope you keep reading.

It's impossible to fully separate politics from open source software, as much as it is from public health policy, education, or nature conservancy. The roots of the free and open source movement are very political, as convenient as it is for us to ignore this fact. The economics, competing philosophies, and moral paradoxes of the material world often surface in software development. Occasionally, the protests, reforms, and conflicts of the material world do too. This is the interconnectivity of our modern ecosystem again. Open source licensing is not spared in all of this—no—at times it's a flashpoint.

When Lerna relicensed, I was participatory as a user. I instructed my team to pin the dependency explicitly to before the relicensing, while it was still MIT. At the time, I didn't have the wisdom or foresight to consider the impact of this license change on compliance, enforceability, or automated tooling, let alone its morality. This book's concepts on the topic were not yet formed.

My concern was stability. Lerna is one of hundreds of tools we use. We, like many teams, don't have a lot of prioritized time to retool or refactor unplanned disruptions. The impact of this license change was not felt by the corporations deny-listed in the new license, but rather on the teams and individuals juggling their own work/life balance. We and many others were collateral damage. It's a near-impossible value exchange to achieve without harming one constituency or another.

I waver...and think...that is in part the point. I empathize with the position of the maintainer but disagree on the approach. Aside from the poor execution of the change, I find it uniquely ineffective due to the systems within which open source software operates. People just forked it, atop the technicality that maintainers had no legal standing to relicense in the first place. The result seemed net-negative—the spectrum of engagement overreached.

Is a tool maker culpable for the violent actions of its users? How culpable is the manufacturer of a screwdriver if a murderer uses it against someone? What if the screwdriver is used to construct a combat drone? Here I see degrees of separation and intent that we must account for. How many degrees is necessary to sleep at night? I don't know. Does the tactile nature of the hardware matter? Moving to software, is an open source maintainer liable for the actions of their derivatives? This wasn't an AI model being trained to identify airstrike targets. That approaches simpler clarity—*the intent is to harm*. But what about an open source GPS system added to a missile guidance system? This isn't hyperbole.

The value exchange of toolmaker is precarious, whether we're assessing intent, impact, or liability. The JSHint essays, aptly named *You May Finally Use JSHint for Evil*,[178] document the paradox of licensing neutrality. Look for more on that in the Resources section of the book. The EU Cyber Resiliency Act "upends this value exchange and tries to place liability upstream in the open source project for downstream security issues."[179] Does the maintainer have agency to act here? Lerna maintainers answered: yes. The Open Source Definition would have us believe this is a simple answer: no. But life is far more complicated than that. We all need to draw our own conclusions and act according to our values and invested energy.

Open source as protest. Open source as progress.

Change is slow and haphazard, yet it's encouraging to see to the arc of time bend toward better licenses. The Ethical Source movement inverts the established neutral structure of open source licenses and instead asserts basic human rights. The Organization for Ethical Source advocates for licenses defining value exchanges based on accessibility, safety, and privacy. They use the legal permissions, conditions, and limitations we defined earlier to communicate these values.[180] One license they promote, the Hippocratic License, features a license builder feature that allows maintainers to adopt modules that reflect their specific values, such as forbidding fossil fuel companies, militaries, or repressive governments from using the software.[181] The Open Source Definition states that an open source license must not discriminate against persons, groups, or fields of endeavor. These two world-views of software creation and distribution oppose one another. Clear differences of opinion and approach, and a lot of gray area individuals, teams, and companies must negotiate.

I've had a lot of time to reflect on the Lerna relicensing. Over time, more information and lived experience has helped shape a nuanced perspective on the topic across legal, moral, and feasible dimensions. My position is vulnerable, evolving, and sometimes I feel like a weathervane whipping in the changing winds. I can empathize with any position with adequate honesty. As time goes on, however, and I see the world needing more light, I cannot help but wonder if the Lerna event should be more inspiration than cautionary tale.

You won't see me use the word "neutral" again. I deeply understand its merits, but cannot ignore its cowardice. The world needs more licenses like the Hippocratic License or the Functional Source License. The moral certainty they provide is what we need, even at the cost of implementation ambiguity. They accurately portray the software landscape as complex and unequal—yet are working to right it. They come closer to justice than any libertarian ideal open source does. I believe we must equally work within our roles. We can educate and advocate for full acknowledgement that evolving software licensing carries risk worth navigating, even if that takes more mental energy and follow-through. I've since become plenty comfortable accepting risk using ambiguous licenses. The world is complicated and we must live within it.

And even with these thoughts, I also have a duty to explain the environment today: large companies blanket-ban the use of projects containing certain open source licenses such as the Unlicense and the WTFPL. Even more economically-focused licenses, like the SSPL or BUSL. Not out of avarice, indifference, or malignant intent, but rather, lack of established legal clarity and risk-aversion. There simply isn't enough case law out there. Automated license compliance tooling can dissuade or prohibit teams and individuals from using your software. Would-be users *and contributors* are locked out of engaging with your software, not because of misaligned ideals, but rather some simple `if-then` checks. It's worth noting, then, that depending on your designed value exchange as a maintainer and licenser, this may be a bug or a feature! Since businesses rely on licensing jurisprudence, weigh the likely outcomes with your ethics. Do you focus on blocking the corporation or empowering the humans making a living there? How complicit is an employee in the actions of their employer? I can slam dunk that argument in most cases, but others give me pause.

Let me emphasize: this isn't a rejection of ethical principles; it's a clear-eyed look at where we stand today. Tech capitalism thrives off the labor of open source maintainers. It's a bleak perspective with evidence to back it up. One could see these newer licenses as a response to, a protest of, the imbalanced and exploitative nature of open source consumption. And they'd be right to be loud about it. Right to attract attention. Historical and recent acts of protest draw media attention to causes of women's suffrage, the Vietnam War, racial justice, and the climate crisis. (FIG 4.9) Do they amplify the message or distract from it?

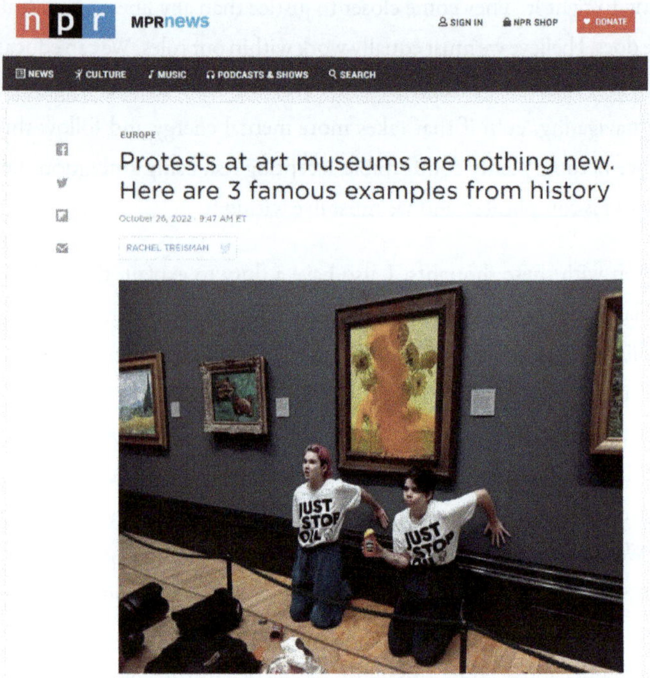

FIG 4.9: Climate activists generated a lot of attention by targeting famous artwork. This may seem tangential to their aims of curbing reliance on fossil fuels, but the message is amplified by the means. Photo from Just Stop Oil/Reuters via https://www.npr.org/2022/10/26/1131377513/museum-protests-famous-artworks-history

In a recent streak of protests, artwork was attacked to juxtapose hypocrisy or highlight looming worldwide crises.[182] When protesting, we shouldn't conform to the medium of the oppressor. But there are practical limits within some systems. Do we award any credit for only attacking artwork under glass? Here we can observe calculated protest, a balance of intent and impact. How different is it from the open source protest I've described earlier? In some ways it's similar to the pursuit of just, humane, and equitable software licensing. In some ways it's not. I try to listen, continually learn, and to paint a broad picture of it all here as I've seen it play out within open source communities.

So after criticizing the methods of others, I should suggest my own. What would modulated protest look like? I posit that you can push for change while still doing right by your users. Microsoft acquired GitHub not long after the Lerna relicensing event took place, renewing calls for both companies to drop ICE. As far as I know—they haven't. I have nuanced perspectives on GitHub the product and company. It's undoubtedly been a force for good in many projects and communities—and has made public, positive, reflective changes over the years—while also having organizational, legal, cultural, and business conflicts that don't always make them the paragon of free and open source stewardship. To be fair, no company is perfect here and it's naïve to think otherwise. I hope I've struck that balance in the pages of this book. We can be participants in an ecosystem while advocating for its improvement. I still use GitHub every day, optimistically, with eyes wide open, and will continue to do so. The Software Freedom Conservancy acknowledges this tight-rope walk. They demand change, for the very reasons Lerna did, as well as unresolved licensing and ethical questions related to the AI-powered, publicly trained Copilot feature.[183] They encourage others to raise their voices, and empower them along a spectrum of engagement that includes migration, boycott, and declaration of GitHub use under protest.[184] These are all more pro-community and don't cause collateral damage to unwitting users. And of course, petitions for "redress of grievance" to your elected representatives is enshrined in another important file.

Where to learn more

That was a lot. Licensing is a prism that reveals a surprising number of serious topics. I hope I didn't discourage you from supporting existing projects or launching your own open source efforts. The connections we can build are simply too strong to ignore. I have no doubt you'll help further the conversation if you choose.

For new work you release yourself—all my previous advice holds. For new work you wish to open source while on-the-clock at an employer, consult your OSPO, lawyer, or Legal team before starting an endeavor. Read GitHub's *The*

Legal Side of Open Source[185] as a gateway into so many more subjects. I've also started diving into Amanda Brock's edited book, *Open Source Law, Policy & Practice.*[186]

Let your license help communicate the value exchanges you expect users and contributors to engage in. If you don't want a weapons manufacturer or a cloud-based competitor using your software, that's entirely up to you. But know these licenses don't satisfy the strictest definitions of open source licensing. Only you can determine if that's alright. Your license is but one part of a complete communications and governance plan that should cover all of the public-facing work your project creates. The license can set up the legal skeleton, but what you do within it is the beating heart that will define your project to the community.

CHANGELOG

The third fundamental file is the `CHANGELOG`. It documents how the project evolves over time. Change can come in many forms, from a `README` bulleted list to blog posts with migration guides, automated code alterations, and video walkthroughs. The change management of your project is another intentionally designed value exchange, and one of the most critical. Once someone starts using your code, they become in part dependent on it. This is a risk to them and a responsibility for you. Responsibly managing and communicating change is the least you can do.

Most projects stick to the convention of defining a release-worthy milestone via a semver-compliant Git tag. When people are evaluating or using your software, they will look to the stability of release as a differentiator. How often does your API change? Do you provide step-by-step guidance when it does? How effectively you communicate version-to-version changes can save them hours and hours of time over the lifespan of their engagement. In the thick of a dependency updates ticket, or when skimming a Dependabot summary, engineers will be making a beeline to your `CHANGELOG` file or releases page to determine if they are impacted.

`git commit`-ment

How you disclose what's changed is up to you, but make it brief and more relevant to end-users than contributors, if you have to choose. Why? Because contributors are more comfortable and motivated to look deeper into a project's Git commit history. When time is tight, optimize for the other developer. Future you and future me will thank you. A year from now, you will be a stranger to this code too.

A lot of tooling and techniques seek to document change at the `git commit` level. Yes, atomic commits[187] are fantastic to review or revert, but require a lot of mental energy to craft. The result is adoption of specifications like the conventional commit syntax.[188] Tooling can automatically pick up these commits and do things like update a `CHANGELOG`, increment your documented semver, even `git tag` your codebase. This is powerful stuff, but easy to mess up and easier still for your contributors. You don't want to nitpick an external contributor's process—they are already going out of their way to help you. You need to meet them where they are. Often, development looks like this:

```
feat: WIP date refactor
does this work?
more WIP
forgot a file
fix lint in CI
```

Accept and sidestep these problems by advocating for basic Git hygiene:

- **Branch by default.** Create a safe, shareable space for each new body of work. You can work on as many of them as needed, independent of one another, and use tooling like GitHub pull requests to test and review them carefully.

- **Keep all change small.** Limit the scope of work to a single complete feature or fix. This is what we mean by making atomic changes. See something unrelated in need of sprucing up? Resist the urge to sprinkle it in. You want potential reviewers to have a straightforward review, both in coherence and in brevity. Sometimes small change is impossible—the work loses clarity if broken up more. And with those large sizes come big commit lists tempting you to rewrite. There is a time and place to revise Git history, but I'd argue they aren't normal days in the office.

- **Set the mood.** Use brief, second-person imperative commands, where the subject, *you*, is implied. (Like that last sentence.) Even if you don't end up using tooling to read these commits, writing them this way focuses the effort again toward the impact of the change. This can help when reviewing code or history.

- **Relax.** For as much as it matters, it also doesn't. Fretting over perfect history, rebasing, submodules, worktrees, it's enough to overwhelm. Approach the depth of the tool as ready, if at all.

A compromise approach to the rigidity of a prescribed commit structure is to squash-and-merge all commits upon merge.[189] This gives you as a maintainer control over the final commit message content, while allowing for commit-level leeway during feature development. See the Resources section for more Git recommendations.

Releases and publication

Sites like GitHub and GitLab package up sets of change as a "release." These platforms allow you to associate some long-form markdown text to a Git tag. If I cannot find a `CHANGELOG` file, I look for releases. Releases also can contain project artifacts like binaries or archive files for end users to directly interact with.

If you intend for your project to be available on the public npm registry for users to download, you need to publish the artifacts too.[190] Keep your Git repository, `CHANGELOG`, releases, and available registry packages in synchronicity whenever possible. Published code omitted from a repository is lazy at best and suspicious at worst—sometimes evidence of ill intent like malware. Unreleased code in a repository is less accessible to others, diminishing its reach.

Remember that in most circumstances, people encountering your project's release documentation or artifacts will either be 1) in a hurry to upgrade with as little effort as possible, or 2) looking to get up and running as quickly as possible. In both cases, clear, concise accounting of changes creates a healthy exchange of expectations.

CODE_OF_CONDUCT

Adding a `CODE_OF_CONDUCT` file to your repository or website makes clear to everyone engaged with your project what the rules are. It defines what is and isn't acceptable behavior within what media, who is covered by the policy, enforcement guidelines, how to report problems, and what happens after a concern is raised. The preamble of the popular Contributor Covenant code of conduct is representative of the intent and the stakes[191]:

> *We as members, contributors, and leaders pledge to make participation in our community a harassment-free experience for everyone, regardless of age, body size, visible or invisible disability, ethnicity, sex characteristics, gender identity and expression, level of experience, education, socio-economic status, nationality, personal appearance, race, caste, color, religion, or sexual identity and orientation.*
> *We pledge to act and interact in ways that contribute to an open, welcoming, diverse, inclusive, and healthy community.*

GitHub makes adding a code of conduct a wizard-like process,[192] allowing users to choose between popular codes of conduct and add contact info for reports. Having a code of conduct on file results in its display on the repository home page, new issue page, discussion page, and elsewhere. It factors directly into the repository's "community standards" insight metric. This makes the process about as simple as it can get. But what happens next matters more.

Beyond a code of conduct is governance

A code of conduct makes expectations public and clear for all parties, but it's the first step in a holistic effort to create an inclusive experience. With a code of conduct in place, it's vital that maintainers are trained and prepared to enforce it. Adopting a governance plan beyond the initial reporting mechanism creates the follow-through that yields real safety.

For example, the Node.js project contains a two-thousand-word moderation policy[193] that outlines who the code of conduct applies to (everyone), emphasizes the opportunity for all parties to assume self-corrective positive intent, and guides maintainers and users through what should happen when a code of conduct violation is reported. It includes four means to request moderation, some near-anonymous or out of band from where the abuse is happening. The policy also details escalation of moderation and how to make changes to the policy itself. Important also is what to do in the event of a moderation dispute. This fidelity was the result of hard-learned lessons, incrementally applied to become a more inclusive, clear, introspective, and objective process. Projects that invest in this work are communicating that people matter more than code, and that should be applauded.

Everyone benefits from a code of conduct

Why bother with a code of conduct? In all my encounters, it is only men that ask that question. They are telling on themselves. They state that you don't need a code of conduct for other places: the corner pub, college classroom, job site, or grocery store. And yet life would probably be better with one, and is! Even a cursory search brings us to employee handbooks, collegiate codes of conduct, and trespass laws . Shoutout to the University of Minnesota code of conduct,[194] which defines 13 different enforcement steps, and outlines hearing and appeals processes. The revision history dates back to 1970!

Luckily, we as a *profession* are catching up to basic rules of decency and inclusion. That means many have come before us and taken a stand. People like Sage Sharp, who demanded more from the Linux kernel community.[195] A community that briefly enshrined an inadequate "code of conflict."[196] Sage, and others, have seen the rampant sexism, inequality, abuse, and favoritism of die-hard hacker culture, and rejected it for something better. That isn't controversial—it's civilization knocking on the doorstep. Fortunately, you can use this as a filter. If someone cannot handle basic decency, you know what to do now. History is full of important contributions made by unsavory persons. We don't need to glorify their contributions despite their failings. We, and those curmudgeons, must demand more of our collective selves. Tech and open source are speedrunning the same cycle at play across broader society.

As these changes play out, "cultural" pushback from an old-guard of personalities and their followers surfaces within discussion forums, comment threads, and even during board elections for the Open Source Initiative standards body. Progress is apparently an existential threat to them. They tantrum and complain of censorship, lashing out at would-be allies and contributors, if only they saw it that way. The vitriol is sign that ever-increasing inclusion is working. There is no doubt much more work to be done, but improvements arrive across the industry, in big and small ways. As Nathan Schneider put it: "If the early architects of open source feel threatened by a new generation demanding more, that is probably a sign of progress."[197]

An in-person industry event, and surely a thing as simple as some message boards around a code repository, can get this right. Jen Simmons, in considering a code of conduct to a conference, uses the analogy of fire code and building safety. Within a discussion on the blog *CSS-Tricks*, she wrote[198]:

> *Many of these conferences are held in hotels. The hotel itself has a plan about what to do if there's a fire in the building. There are laws, in fact, about the hotel and fire safety. The fire marshal visits periodically to make sure those laws are being obeyed. What is considered safe is figured out ahead of time. When and how to evacuate the building is figured out ahead of time. What equipment is needed (detectors, sprinklers, standpipes) is figured out ahead of time. It's not figured out during a fire. It's not dependent on who's there, or why a fire was started. It's decided long before, when there is no problem in sight, so a problem does happen, people have a clue about what to do in the moment.*

These policies are designed and distributed *before* something goes wrong. As Simmons says, it would be unthinkable to devise an evacuation gameplan during an emergency. It's why we practice fire drills. What we do now to prepare for difficulty makes facing those situations easier, because we have a shared understanding, a written record, and muscle memory to spur our actions.

I almost omitted a code of conduct as a fundamental, but not for lack of importance. Rather, most projects start with a non-existent or closed community. Maybe a population of one. Given everything else going through your mind when open sourcing your fledgling project, moderating a community might feel overwhelming. But it costs you nothing to adopt an inclusive mindset, and a lot of help is out there. Sameera Kapila says within *Inclusive Design Communities*[199] that "if everyone takes even one or two of these small steps toward inclusivity, I believe we'll see change."

Start Simple

Maybe you start with a small set of files you want to get off your local machine. Maybe you want to share code with your network. As Brad Frost says in a long-pinned social media post: "Work hard. Don't be an asshole. Share what you know." That's succinct! But often we need more guardrails. Create clear expectations with these four fundamental files, README, LICENSE, CHANGELOG, and CODE_OF_CONDUCT—you won't regret it. One day, you might just find yourself in the company of friends and strangers. These folks are allies when respected and welcomed. They'll tell you in word and deed what else the project needs. Listen. Next chapter will explore how to do just that.

Chapter Five

Community Investment

When starting any task, it's instructive to look at the finished product as a reference. We see plenty of examples on GitHub of mature projects that seem unabashedly complex. The intricacy of their release process, the thoroughness of their test suite, the size of their engaged community, the depth of their documentation—it all adds up to a mountain of completeness you couldn't possibly emulate in your own open source work. Maintaining an open source project can feel a lot like drawing an owl. (FIG 5.1)[200]

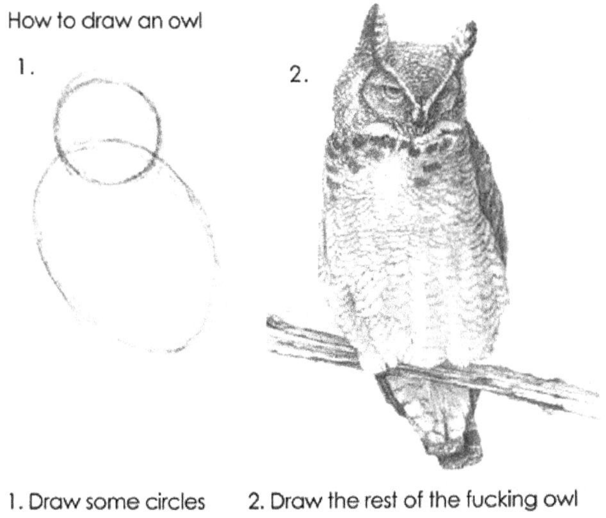

FIG 5.1: Open source work can feel like this. "How to draw an owl: Lessons in becoming an expert in a couple of easy steps" by Cea. is licensed under CC BY 2.0.

How so? These examples are intimidating and hard to imagine as the result of steadfast improvement. It's important to highlight that none of these projects started this way. They started as ill-defined, incomplete circles, not owls. Hundreds of intermediate steps combined to build what you see today. Dan Mall reminds us of this within any creative process when he tells us that "it's easier to revise than create."[201] The project was improved day by day, commit by commit, contribution by contribution.

You simply cannot do it all at once. That's why last chapter focused narrowly on fundamental concepts that serve your project well in its most immature or selfish state: useful to you, future you, and what I'd call motivated masochists, or maybe nicer labeled early adopters. A `README`, `LICENSE`, `CHANGELOG`, and `CODE_OF_CONDUCT` are the least you can do to document, protect, and communicate the seedling of your project.

Some projects may not meet or eclipse this moment, and that is fine. But some grow into more. Sustained work attracts more people, each with different needs, priorities, and constraints competing for attention. If you aren't careful in your designed engagement, the very network you've worked to build can work against you. Let's outline a few areas to invest in.

Documentation

The more software you write, the more you realize that code is only part of the journey. As the surface area, nuance, or use cases of a project grow you naturally need to document it. *Do I need to document the release process?* (The answer is always yes, by the way.) The corners you cut, the fundamentals you neglect, become untenable omissions when sharing the project more widely. When we leave town for a few days, my son leaves a note for my mother so she knows how to care for the pets. Even down to the lighting care for the fish tank. Without this, she'd be left to guess. Think of documentation as a first-class artifact of the product—on as equal footing as the code itself.

But it's not enough to merely prioritize documentation equal to the code. The most thorough docs in the world won't help your users find what they are looking for if they aren't well-structured or approachable. What if you want to use your docs to show more than tell, or reach broader audiences?

Isolated examples

An exhaustive accounting of API methods and options has its place. A time will come when you need to see the exact response object schema from an endpoint, or the layout of parameters in a function. But you can diversify the method of learning this content with a little extra effort. For example, the MDN documentation for the CSS property `justify-content` (FIG 5.2)[202] not only defines the syntax and links to the specification and browser support tables, but front and center is an interactive playground feature.

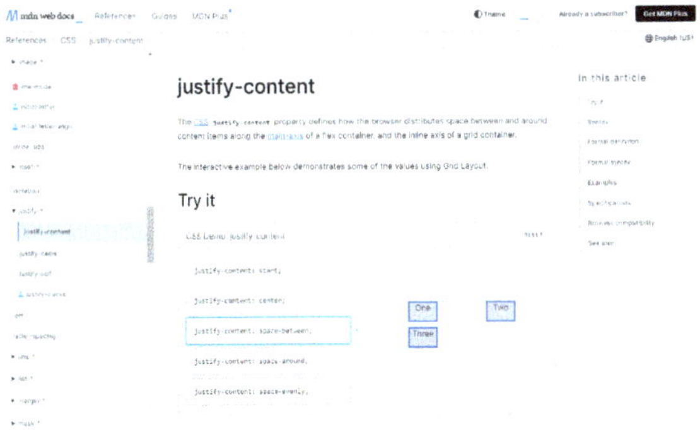

FIG 5.2: The justify-content CSS property on MDN with a demo among the first content.

The documentation itself demonstrates different property values, from centering content to spacing them evenly, etc. This visual is just what someone might need to jog their memory or grasp a concept when they are in the midst of another task. And if they need to dive deeper, they are free to explore further.

Chris Coyier's oft-bookmarked *CSS-Tricks* article "A Complete Guide to Flexbox"[203] includes an embedded editor of the online CodePen tool,[204] letting visitors fiddle with HTML and CSS and see real-time layout changes. Open source projects might write their own sandbox or REPL (Read, Eval, Print, Loop) tool, like Babel's, showing how different configuration alters your transpiled JavaScript input. (FIG 5.3)

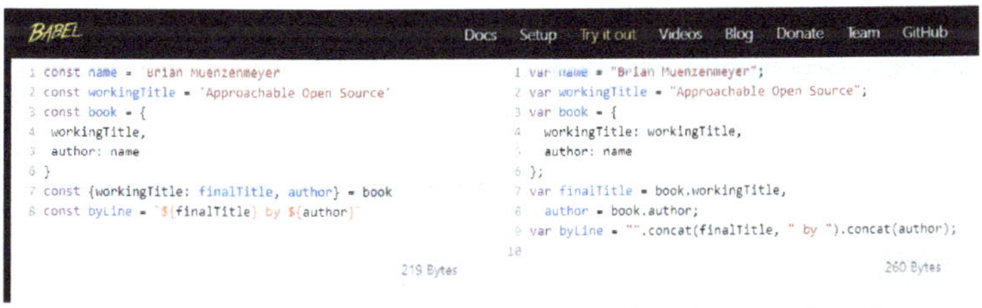

FIG 5.3: The Babel REPL allowing users to try the software online. Input on the left is sent through the library and output on the right. Explore options, all without installing the tool.

A tool like this available without the need to install it yourself has immense impact on the approachability of the project. Users can learn about your project without having to invest their own time, filesystem, or infrastructure to do so. It might serve as an excellent end-to-end test as well, keeping your input and output aligned to expectations in their purest form. I note, in my typical cautionary way, that these features incur maintenance and stability risks. They become products themselves, far beyond a `README`.

Search

It's hard to deny the simple utility of `cmd` or `ctrl + f` to look for something within a document. Some open source projects really lean into this concept, like lodash[205] making massive single-page files defining their entire API. The static site generator Eleventy adopts a similar low-key but effective search strategy,[206] passing search terms to an external search provider subset to their own website. The result is a target query like `date formatting site:www.11ty.dev` which search engines are great at indexing and filtering.

But when your information architecture places content across files or pages, and you want a more integrated experience, built-in search adds increasing value. Many open source projects apply for free search capabilities from the easy-to-integrate Algolia provider. (FIG 5.4)[207]

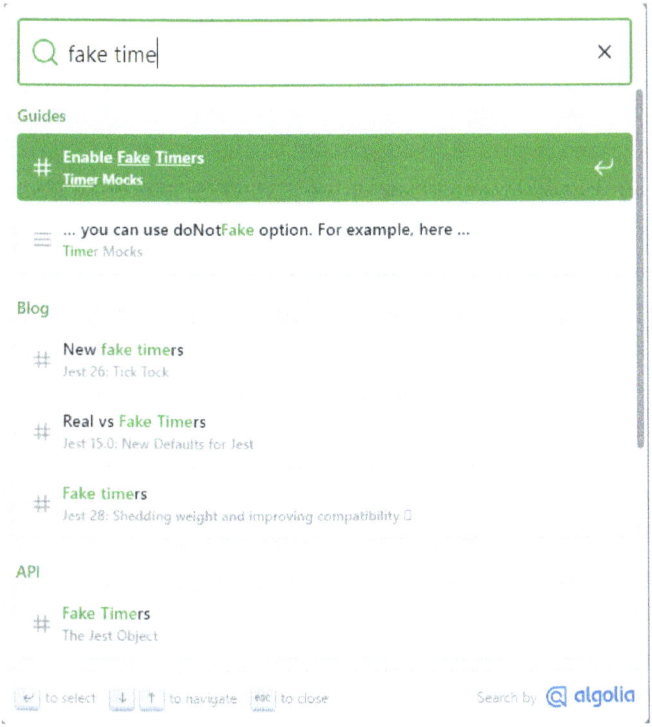

FIG 5.4: Search results powered by Algolia, of the jest testing framework. Be careful what you index, however. Most people probably don't want a blog post from 2016 before the actual current API info.

There will be pressure as you start to accrue API documentation, guides, tutorials, blog posts, meta files, and everything else. Search can act as a safety valve. A well-designed information architecture is possible with time and effort, but search can circumvent or mitigate sprawl. A well-crafted search experience can reach across all semantic boundaries of content to help a user find what they are looking for. You'll find that some users prefer each method of content discovery. Again, this extra effort represents more time spent away from your real product, but can pay dividends to you and your adopters.

Internationalization

A steady refrain here has been that as your project grows, you'll attract more and more users from diverse backgrounds. English should not be the assumed, singular, mode of communication on the World Wide Web. While around sixty percent of all web content is in English,[208] Hootsuite reported that thirty-two percent of Internet users translated text on a website within a single week.[209] In the technology and open source sector, content is skewed toward English,[210] even as the global community of engineers and open source participants grows. It's hard to conclude an appetite for multilingual content is non-existent.

Still, many systems require English, such as the Common Vulnerabilities and Exposures (CVE) spec.[211] And some folks even suggest English-only or English-default technical content is more precise and accessible than fragmented and poorly translated variants.[212] That link points to a blog post where the author self-describes this position as an "ugly American" viewpoint. This complex tug-of-war isn't limited to open source or tech. The Dutch government is proposing a limit on English-language university curricula popular with foreign students.[213] This highlights the continual balance sought between intellectual openness, talent competition, and national economic pressures.

Me? I'm privileged enough to have English as my first language. I've lived and worked within a system that benefits me across many dimensions, including language. I haven't felt the sting of having to search for an error message that looks doubly unfamiliar to me. I respect the notion that if thirty-three other language communities work to translate the semver docs,[214] there is content there of value.

While it might feel presumptuous to *plan* for translation—it doesn't hurt to at least consider this eventuality when you feel your project is on the cusp of broader relevance. Users interested in translated content will approach you, and an early decision may make translation easier. Factor in your internationalization posture as a dimension in the `README` vs. website decision. Adding internationalization to a custom website could mean a complete rewrite, versus choosing something that has internationalization routing built in.

When to break the rules

If you ever catch yourself or your team stating that you will write the documentation later, stop and reflect on if that's the right prioritization. Very rarely is this an intelligent deferment. Like doing the dishes, it gets harder and harder to catch up the longer you postpone. When your solution is still in extremely rapid API development or shifting approach, you have cause to wait. Likewise, there are things you may want to document, but perhaps not make broadly available to the public. Onboarding guidance for mentors, community moderation proceedings, and security vulnerabilities may warrant discretion. Private documentation is still better than no documentation.

Projects that stand the test of time

The *Bjarkemål* is an Old Norse poem I studied in college. [215] It tells the tale of Rolf Krake and his retainers as they awake to battle a numerically superior foe. Stories, songs, and poems were essential oral tradition, but many are lost to time, including most of the *Bjarkemål*. The *legacy* of this poem, however, is what really excited me, even as the source material no longer exists.

- There are accounts of this poem being sung by bards *contemporaneously*, used to amp up the troops of another battle some 30 years later.
- There is a revisionist push to tie the poem to a newly-sainted king dying in battle. Or was he embarrassingly ambushed, even murdered by his own men?
- A Latin translation exists from the author Saxo Grammaticus 200 years later, but it's widely believed to contain large embellishments, additions, and plagiarized themes from the *Aeneid*.

What a curious mess! It's both heartbreaking and fascinating that so much speculation and creative license can stem from one undocumented, oral story. How much more do we get wrong, or lose in translation?

In contrast to the *Bjarkemål*, the world has the *Tripitaka Koreana* woodblocks housed within the Temple of Haeinsa in South Korea.[216] (FIG 5.5) These 81,258 wood carvings date back to the time of Saxo, and have been perfectly preserved using techniques invented and mastered there. Within each block are hand-carved characters representing Buddhist scripture. In all, they store fifty-two million characters of data, meticulously preserved for eight centuries. The world may not have ever seen a more thorough, accurate, and lasting documentation effort.

FIG 5.5: Tripitaka Koreana woodblocks meticulously preserved for eight hundred years. Photo from The Library: A World History by James Campbell and photographer Will Pryce

These two works, the latter a success and the former quite lossy, align nicely with GitHub's goals within the Arctic Code Vault program[217] to preserve open source information for the next one thousand years. In 2020, GitHub placed twenty-one terabytes of repository data within an adapted mine at Svalbard, a Norwegian island archipelago near the North Pole. The site is geologically stable, naturally temperature controlled, and located a mile from the Global Seed Vault.[218] Let's hope our `README`s are read by our future maintainers.

Here and now, our open source projects exist on a much shorter timescale. Even the venerable Express web server project isn't old enough to drive yet. Git is almost old enough to vote in the United States. curl has a car payment, and maybe a mortgage if it's lucky. These projects persist on the resilience of their communities. The Node.js team celebrated ten years of the project by holding planning sessions to steer the project into the next ten years. Their open governance almost guarantees success. Implementing our fundamentals within a project preserves the intent necessary to understand the work. Recording the instructions defines the steps required to execute software. Once released, most software never really dies. It lingers on CDNs, package registries, in dormant code repositories, a half-life we've barely begun to understand or measure at scale.

We are privileged to be inventing our digital norms right now, a continuous chain of decisions dating back a single lifetime. A *symmathesy*! As long as software holds true to the mantra of "don't break the Web," we will be able to look back to yesterday. The general availability of software under the semver model of versioning allows us to march forward with confidence while preserving the past. It's delightful to pick up an old project and see it run, even over ten years later. I recently hosted my first jQuery plugin, CompassRose,[219] as a sort of time capsule. The code arranges and manipulates DOM elements along a unit circle. (FIG 5.6) It's more toy than tech at this point, but its ease of use even now is a testament to the availability of software long past its shelf-life's expiry.

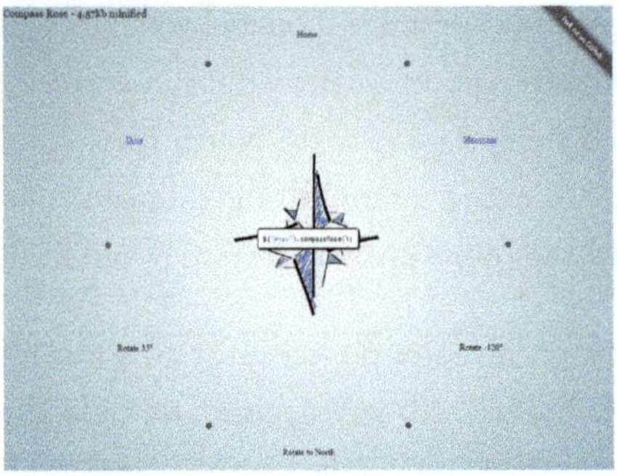

FIG 5.6: The still-working demo site for my first jQuery plugin, circa 2012.

When working within the Node.js ecosystem, include a lockfile, the package manager used, and the required Node.js version. This provides hints to current and future users how to pick up the project. For Chapter 3, I locally cloned and ran the graph software necessary to generate the dependency comparisons of Next.js and Create React App. It had a three year old `package-lock.json` lockfile, which resolves and helps install 546 separate packages, and "just works." Ten or three years of entropy without incident isn't eight hundred, but I'll take it.

Community As Product

The atmosphere we create around our open source product is an important determinant of its progress and longevity. This community can be intentionally designed—it's not enough to put something out into the world and expect overnight success. It will take sustained effort. Starting something is a lot easier than maintaining it, observed within the Linux Foundation census.[220] Maintainers often leave the projects they begin. Whether the project lives or dies is a litmus test. How self-sustaining has the community become? More contributors create inertia that keep a project moving. With proper planning, governance, and safety, contributors plot a course, direct traffic, clear a path, push in unison, and help one another rather than play tug of war or fret over direction. To create projects to stand the test of time, you need to design a space that is welcoming to all. You need to invest development time into creating features that create connection. You need to communicate transparently in every aspect of the project.

And lastly, you need to listen. Open source, at its best, is about creating something larger than yourself. That only happens when you let others in. We can interact with one another as both user and maintainer like an episode of the Great British Bake Off, not Master Chef. The relationship can be supportive and respectful, not adversarial or competitive.

Creating avenues for feedback and communication

Feedback is like dust. It's all over, the result of every action, but only visible if we pause to look at what's settled or observe that rare sunbeam of clarity. Often you'll notice feedback when things go wrong. The Harvard Business Review found that the most effective teams exhibited a ratio of over five compliments for every criticism,[221] a ratio unlikely to be found on even the most successful open source projects. This can be frustrating—you get used to it—but if you are persistent, lucky, and exercise empathy it never quite stops pushing you to improve.

Feedback is a moment of engagement, a real-time value exchange where user and maintainer can level with one another. User to user connections leapfrog you as maintainer to accelerate learning and coalesce effort around the known boundaries of your program. You might find out that an inadequate setting or shortcoming on your project has sprung up a cottage industry of workarounds. Don't be offended—do your best to enter into these exchanges with an open mind, civility, and preparedness of action.

Consider ways to connect users of your software. Whether you like it or not, you need not even be a party to the conversation. Crowd-sourced help and troubleshooting offer more ways for your users to connect, leading to a novel approach or resolution of a problem, collaboration, and collective contribution.

It humanizes the value exchange farther than a **README** ever could, and if individuals want to, can open up doors. Fostering communication and networking beyond you as the hub expands the engagement of others toward shared ownership.

Sites like Reddit allow for subcommunities to congregate around specific topics. Perhaps your project is large enough to warrant one, like the Inkscape[222] or Eleventy[223] communities. Question and answer site StackOverflow lets anyone create topic-tags[224] to focus, filter, and follow conversations. (FIG 5.7) Users can even subscribe to an RSS feed.

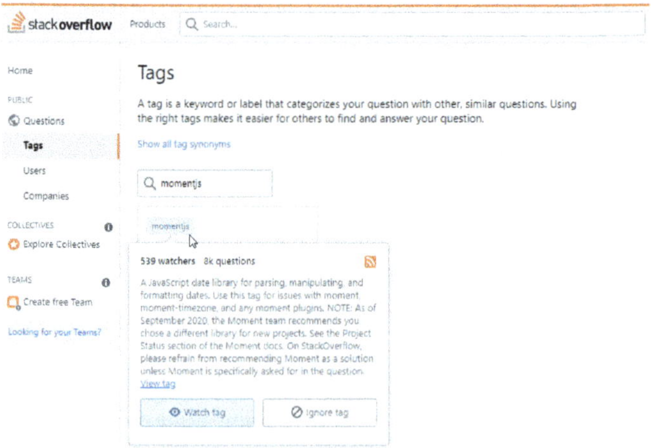

FIG 5.7: Stack Overflow tag for Moment.js library. 539 people are watching for new questions for this library.

Beyond an issue log are many chat programs that look to introduce real-time communication into a project. Discord, Slack, Gitter, and IRC all have features that allow users to message with one another. They offer special-interest channels, direct messaging, and creature-comforts like bots, emoji, and personalization. Often they have mobile clients that make access a swipe away. Good for idle-time triage, and again ripe for too much intrusion—you decide! Your project's chat footprint becomes another area where the code of conduct applies. Ensure this new avenue of engagement remains a safe place for all. Set clear expectations, and look for responsible members to help moderate these spaces. Also be careful not to slide into a state where the only communication occurs through these channels—you still want your project to be accessible to those that don't use chat.

If you want something with better longevity and discovery than a chat firehose, turning to tried and true online methods like an email or forum are hard to beat. Mailing lists have a surprising degree of activity within older projects. Bulletin board software still exists out there too—even if you haven't visited one since the likes of phpBB dominated the web's early Q&A communities. Check out some options in the Resources section. Open source projects can often apply for a free access tier to tools like these. The Vivaldi browser project operates a forum[225] complete with large topic areas and sixteen language-specific areas to meet their users across the globe. This can be a really nice extension for your project if you expect to interact with non-developer users. The need for an account to interact beyond browsing code repositories might be off-putting or confusing to your audience.

For developer-centric communication, GitHub has a discussions feature that integrates into repositories or organizations.[226] It's convenient as a maintainer to have a single UI to go to for all things related to your project. The tight-coupling of a discussion to other aspects of your codebase is helpful, too. If someone opens an issue that is vague, general, or looking for guidance, convert the issue into a discussion. This focuses coding-related efforts within the issue log and community and user-facing help within discussions. Each can have their own participants, focus areas, and engagement patterns. A discussion forum like this is a nice entry point for users looking to contribute more to the project too, lending their expertise to others.

Seek out users where they are and you'll lower the barrier to engage. It's an investment—one that can create new pathways for the project. Make sure you do this on your own terms.

Network overload

My father once told me a story about priorities, in which his boss had asked him to do one too many things at a time. They sat down to talk about it. My dad presented a sticky note for each task deemed a high priority to get done. He slowly laid them all out on the desk. It was probably a bit tactless, but he's sorta that way. With limited staff and resources, there was obviously too much. If everything is a high priority, nothing is. By the eighth or ninth note, his boss relented. All those other priorities were more important than this new one. He was "managing up" his boss on kanban or agile principles outside the realm of software development.

With a project of increasing size, you may find yourself with a chat presence on Gitter, Slack, Discord, or whatever the future brings. Folks might start chattering on social media and across your inboxes. Notifications can quickly run rampant. The coding platforms we use today generate notifications for nearly every activity on a subscribed repository. I once took a week out of the office and returned to over two-thousand notifications. (FIG 5.8)

FIG 5.8: My notification inbox showing 2007 items to review. I haven't tuned what I am subscribed too well enough.

When many users, automations, and robots converge in your inbox, it can be easy for the important stuff to slip by amidst the volume of distractions. While employing no ill intent, well-meaning actors can inundate maintainers.

Tools that give you control over scope, interval, medium, and intrusiveness help reduce noise. One of the simplest things you can do to reclaim agency over your project is to configure notification settings. Expose yourself to updates under terms that are sustainable and healthy to you. I recommend:

- turning off default, global email notifications
- using the web app or native app intentionally, with as few push notifications as possible
- selectively subscribing to repository actions with high impact (like pull requests or direct mentions)
- filtering notifications of dependent projects to releases only
- using a weekly digest for low-priority notifications, such as Dependabot alerts[227]
- leveraging email headers and CC fields to further filter on urgency[228]

Once you develop a healthy notification protocol, you won't feel as on-edge about any one update. The important stuff rises to the top—the rest can wait. You still need to eat, spend time with loved ones, sleep, and commit to self-care. The "speed of open source" works in your favor too.

Meritocracy...?

The last leg of our journey into community as product is a discussion about meritocracy. It's the albatross of open source thinking and needs to be actively dismantled. I reject the notion that meritocracy—governance based on ability alone—holds any redeeming value in successful, long-lived projects. It's an enticing fairy tale to spin that good ideas can come from anywhere, or that measuring worth based solely on project contribution is a virtue worth codifying.

But enshrining meritocracy above all else fails to look critically at the inequities inherit in open source software, as well as the entire tech industry. White, male, cisgendered viewpoints are the default viewpoint, and carry with them more self-perpetuating weight than other perspectives. What does this even mean, from the Apache Software Foundation[229]:

When the group felt that a person had "earned" the merit to be part of the development community

What context do words like "felt," "merit," and "earned" have here? It sounds awfully subjective, their definitions malleable enough to introduce bias and abuse. The Apache Software Foundation does great work, but this isn't among it. You, *yes you*, don't need permission to exist or someone to tell you that you belong. *The Open Source Way* guidebook,[230] an expansive and thoughtful project itself, devotes a section to inclusion and meritocracy worth looking at[231]:

At first glance, open source seems immune to the multi-pronged problems with diversity and inclusion plaguing technology companies. Open source ecosystems operate online; contributors work across countries, languages, and timezones to power projects that drive technology forward. In some cases, people contribute purely out of love for these projects, not for free beer and stock. Given the prevalence of such altruistic intent, the thinking goes, open source is a true meritocracy, a space where all can thrive regardless of their pronouns or the color of their skin. But if there's one thing that industry advocates love more than meritocracy, it's data. And when the data shows who contributes to the open source ecosystem, it paints a damning picture.

To make projects inclusive, you need to "stop saying you're a meritocracy."[232] You need to create asynchronous opportunities to contribute. You need to exercise grace in shepherding less-than-ideal efforts. You need to stop forcing people to decide between dinner with the family and the late hackathon and happy hour. Intertwining alcohol into your event will push people away looking to avoid risky situations.

Oh, and that damning data? It isn't going away. GitHub's 2017 Open Source Survey[233] reported that a staggering ninety-five percent of respondents were men. Four years later, the State of JavaScript and State of CSS surveys reported sixty percent and seventy percent male respondents, with large numbers deciding not to divulge. Their solicitation and results cause quite the commotion on social media each year, with an understandable backlash and respondent boycott.[234] Does one trust, or want to trust, the results of such a one-sided population? Is this self-fulfilling prophecy? Survey organizers invest a lot of time into these efforts, and yet the results skew male.

The 2021 Diversity, Equity, and Inclusion in Open Source report by the Linux Foundation[235] states that just fourteen percent of participants in open source technology are women. Women, as traditional primary caregivers, have less time to work on open source compared to other duties. Further, Black, Latinx, and Indigenous peoples are "far more likely than White/ Caucasian people to experience exclusionary behaviors in open source communities." One anonymous survey respondent puts it succinctly:

> *The meritocracy model typically assumes that people have a ton of time to invest to prove their 'worth'. People who need to work two jobs to make ends meet don't have this time.*

In the *Model View Culture* article, "The Dehumanizing Myth of Meritocracy", Coraline Ada Ehmke goes further than looking at numbers and instead posits that meritocracy relies on a utilitarian viewpoint of human intellectual output as the strongest currency in the land: "The almost religious devotion to an idealized meritocracy perpetuates a system of elitism and virtual classism. It props up an advantaged minority free to spend its time catering to an idealized community of homogenous consumers."[236]

If any modicum of meritocracy should be salvaged, it's that no one should be discouraged from contributing because they started later than someone else. How do you help achieve this? Your project-related writing should be comprehensive to all aspects of the software development lifecycle. It's not good enough to say the code is self-documenting.

Promote transparency over meritocracy to increase diversity, equity, and inclusion efforts. I feel like this comes close to the spirit of meritocracy without assigning any claims to the output of individuals. Instead, transparency in open source increases access to newcomers by placing all context into the public eye. Don't keep logic, plans, or decisions in your head. Write them down, and make them public. Resist the temptation to solely rely on private message boards, private team threads, or direct messages. Allowing newcomers to browse past work lets them establish as much context as they wish, rapidly. Once they have clearer notions of what the project is doing, where it's come from, and where it's going—they will be prepared to contribute.

Consider context-creating opportunities to be:

- **The issue log:** both your and contributors' intent to change something
- **Contribution guidelines:** covered earlier in the `README` section, outlining how to participate
- **The roadmap:** groups of planned work guiding people eager to help but not sure where to start
- **RFCs:** requests for comment, a public solicitation of feedback for substantial future improvements

You'll note that each of these become collaboration tools and engagement moments themselves. With a proper investment in community, you won't be alone in finding opportunities to further your project.

The faster and more broadly you create context the more approachable your project is. Context drives project velocity, growth, and retention. Investing in clear intentions throughout your work will attract more diversity of thought than leaving your community to guess at what's valuable to the whims of a benevolent dictator for life.[237] Ehmke was interviewed in the Linux DEI report I cited before. She slams the door shut on meritocracy's flawed theory that only the code matters:

The whole notion that our differences don't matter is backwards, our differences are what maximize the potential of tech to do good in the world.

We are not alike. We don't have to pretend we are! We should acknowledge and celebrate our differences. Our experiences are not the same. The contributions we make extend beyond code and create *community*.

Automation

Time is scarce. Themes throughout this book, Roman philosophers, and modern life prove this. Our open source reality is no different. What tasks we can defer to a machine gives us that much more time back to devote to what does matter. I write this as a deluge of AI-powered tools comes to market. I'm not advocating for the machines to write all software, but suggest they become sophisticated tools to be used with care, like robots that perform eye surgery with more precision than a human hand.

I suggest you judiciously automate any task that a robot can accomplish with more determinism than mere mortals. Trimming a cornea—or running the test suites. Pick the tasks that have high frequency, probability of manual error, or have a favorable return on investment. You'll create more consistency and clarity within your product offering too.

Contribution expectations backed by healthy automation remove subjectivity and uncertainty for new committers. Because machines never sleep, contributors can acclimate to the context of the project with faster feedback loops—no waiting for pull request feedback that nits the semicolon usage or points out a test discrepancy. Confidence in release improves too when a process can be codified. Let's explore different ways we can introduce automation into our project.

Coding style

Anything you'd have potential blocking feedback on within a merge request should be a candidate for automation. As a good first step, write down your expectations. Some projects, early in their journey, simply state: "in lieu of a style guide, model contributions off of the existing code." This isn't ideal, but it communicates expectations. Better yet, automate your coding style choices.

Find it within yourself to define limits to what you care about. If we are attracting contributions, we need to bend and not break across so many dimensions. Document and automate the things that matter, and channel the energy Cap Watkins calls "the sliding scale of giving a fuck."[238] The stuff you are really serious about, those last-straw principled stands, you build into your testing and linting suites. Everything else is one of those "yes, and..." moments that improv artists and you as a maintainer must embrace.

`.editorconfig`

Resolve tabs vs. spaces arguments forever (hint: tabs are more accessible[239]) by automating your codebase to use one or the other. The EditorConfig project[240] has deep and default integrations across tools, and lets you define text file options via glob patterns. It only concerns itself with the control-characters within a file—leaving other formatting concerns for more robust tools. Aside from indent style and depth, you can control how new lines are handled. By including this file in your project, (FIG 5.9) you hint to and compel contributors to adhere to your standards. This file helps with cross-platform interoperability too, as a Windows user won't accidentally add invisible characters to the end of every line that can cause conflicts.

```
# top-most EditorConfig file
root = true
# match all files
[*]
indent_style = tab
indent_size = 2
end_of_line = lf
charset = utf-8
trim_trailing_whitespace = false
insert_final_newline = false
```

FIG 5.9: An .editorconfig file can automate away cross-platform and user-dependent whitespace concerns.

Lint

The popular code linter ESLint[241] ensures your code adheres to rules that you configure. Say you never want a `console.log` to show up in a production website. Set a lint rule to enforce it. Or maybe you are particular about the order of your dependency imports—there is a rule for that too. ESLint also has a REPL[242] (remember those from earlier?) so users can play with rules without having to invest in a local installation. With a huge variety of built-in default rules, and the ability to define custom rules, it's safe to say that nearly any spicy programming opinion can become a lint rule. Once defined, you get to decide if lint rules are warnings or errors—each can carry different urgency to resolve within your codebase. A common convention is to run lint checks within a CI build so code on the remote is sanitized before merging.

Lint rules are a perfect vehicle to automate project preferences. Make sure you strike the right balance here and don't fall into pedantry. We want to guide contributors toward the right solutions without throwing up too many barriers along the way. Most editors have plugin support to give a developer real-time feedback as they type or save a file. The goal is that there are no surprises come review time—the value exchange is an explicit communication of expectations, a perhaps costly chore, but with a net-gain in consistency and quality.

Format

A compatriot of the linter is the formatter. A properly automated formatter will ensure the code remains readable. Again, editor integration and CI validation blend to bring the best experience to contributors. The powerhouse formatting tool of the day is Prettier,[243] but any will do. Prettier integrates with ESLint to format the code while conducting quality checks. (FIG 5.10)

FIG 5.10: The eslint and prettier tools emitting code-quality guidance via rules setup in the project.

TypeScript

I'm not certain this qualifies as an automation, but electing to write your project with TypeScript ushers in a new level of scrutiny your contributors and users will appreciate—type-safety. With TypeScript, your code becomes adherent to constraints enforced at build time, not runtime. If your function expects a string and you pass it an integer, it fails to build. If you specify a parameter as required and it's omitted, the build fails. The shape of object properties can be defined and asserted. Power comes not only from your first-party code but also dependencies that ship with their own TypeScript definitions. Editor code-completions allow you to develop with spidey-senses tingling the whole time, telling you what is available and what to watch out for.

The TypeScript language is expansive and nuanced, with a learning curve that applies to maintainers and contributors alike. Luckily, it's also the subject of many courses, cheatsheets, tutorials, and independent guides to help. AI assistants shine here too. I suggest you start small with any TypeScript journey. It's too easy to get a bit lost otherwise. As Sean Wang puts it, "Use types. Not too much. Mostly basic." Don't be afraid or ashamed to relax the TypeScript settings within the `tsconfig.json` file of your project. Strict type-safety can be too intensive an investment, like pursuing too much code coverage. Adding the type `unknown`, `any`, or hinting to TypeScript not to bother with a `@ts-ignore` comment really isn't that big a deal. These are safety valves to be used with intent. The acceptance or denial of this is another thing your project can automate via an ESLint rule.

Tests

Tests are a dimension of your project that benefit from automation. This feels so obvious now I almost bother not to say so. But I have two opposing experiences to highlight.

What testing should and shouldn't be

At one of my first jobs, I remember writing test specification documents. The place was a strict waterfall development IT department attached to a manufacturer. The test document was heavily templated from some long-gone contracting firm, and mirrored the issue reproduction format. You know, the "environment setup, steps to reproduce, expected result, actual result" ideal report.

- Except we wrote them ahead of time in a Word document.
- And then we executed them manually.
- And then we recorded the results.

- And then we wrote our name in a table at the end of the document and the date.
- And then we updated the Gantt chart.
- And then no one ever looked at the test document again.

These manual acceptance tests were of course not automated. Their utility was limited to an instant when the project was evaluated as "done." I hope nothing ever changed after that! At my next job I was blown away by the ability to use JsTestDriver[244] to run automated unit tests against every browser I had installed on my machine (and on the build server). I remember installing more browsers with the explicit intent of achieving greater test coverage, especially in the era of Internet Explorer quirks (let me have my "in my day we walked uphill both ways" moment). It wasn't ideal by today's standards, but it's stuck with me as a night and day difference in testing approach, tooling, and expectations.

Automating our tests helps maintainers and contributors know what contracts matter. We make clear our expectations, not of "done," but "right." Tests are an investment—a choice you make as a maintainer against all other priorities. Tests give us confidence that our code works under explicit circumstances. We align our tests with the API to assert value. Contributors can augment code with certainty that they have not broken existing functionality. The development of Pattern Lab Node was aided greatly by its comprehensive test suite of over six hundred tests.[245] Contributors to the project were able to add functionality with a greater sense that a regression would not occur. A test is a promise that you don't need to remember. These promises pinky-swear via a machine that never tires. No manual intervention required.

Ways to automate testing

Assertions should run during key project events to guide activity toward stable outcomes. Here's a short list of events where tests help, and how they can be automated:

- **Before merge of any new code:** This is the traditional, obvious automation. Tools like GitHub Pull Requests can run tests via status checks.[246]

- **Before production release:** Make sure your code works right before you publish. As part of a scripted workflow, or a tool like `np` that builds a test run in.[247]

- **After local development installation:** npm lifecycle scripts like `postinstall` and `prepare` can help contributors run the tests to verify proper setup.[248]

- **Before commit:** Run tests (or lint, or anything) before committing code to Git with tools like husky.[249]

- **During development:** Most test runners like AVA[250]and Jest[251] have a watch mode that will run tests on any code change. Node.js now has an integrated test command with watch mode too.[252]

Remember, seek to verify the code does what you think it will, within given constraints. I don't recommend chasing a one hundred percent code coverage metric—it's deceiving to think that the code is devoid of bugs with this measure. I've seen teams get caught up in pursuit of too much testing. Back away slowly if someone ever starts selling you on one hundred percent test coverage—they are one hundred percent wasting their time (and yours). Look for confidence in release. Find the right fidelity of tests based on the functionality of the software. For example, a website needs user-centric functional tests. A web server needs memory-optimizing perf benchmarks. A library may suffice with unit tests that check specific input and output.

Supply chain security

Already covered in Chapter 3, automated dependency management tools are built into coding platforms or reasonably easy to integrate. But worth noting here is that as a maintainer of an open source project, you now enter into the complicated food web of dependencies not only as a consumer but as a consumed entity. It's your responsibility to update your own dependencies and distribute them safely to downstream users. Look to ways you can configure the update tooling to work for you, not against you. Most let you set schedules, bundle changes, or ignore worrisome or brittle packages. Skimping on maintenance can put you and your users at risk. Projects that stagnate but become part of dependency chains make for hard-to-resolve, noisy experiences, such as when create-react-app had to pin a disclaimer on their issue log to head off confusion.[253]

The more dependencies you have, the more automated dependency updates may benefit or burden you. Why? It comes down to testing strategy. You of course don't want to be in the business of writing tests for your dependencies. That's their job! But, an automated dependency update needs to be verified nonetheless. Fortunately, your continuous integration system will run the assertions you've configured as part of the pull request. (FIG 5.11) Is that sufficient?

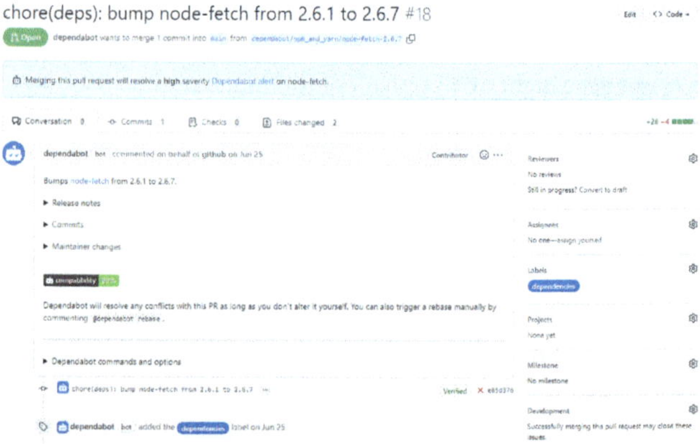

FIG 5.11: Do you have adequate functional tests to know that node-fetch will continue to work?

The breadth and depth of your tests will determine your confidence that the dependency update is safe. Inadequate test coverage, or more appropriately, inadequate testing *strategy*, will force you to accept the changes blindly or revert back to manual testing. For example, if a pull request only updates a dependency, but you explicitly mock that dependency in your test suites, how do you know it's still working? Because of this conundrum, I advocate for functional test suites that exercise your project like a user would, rather than unit tests that only focus on individual parts. Both have their use cases, but only functional tests can build automated dependency update confidence.

Preview

Services exist to show contributors and maintainers how a code change will affect a product. Cloudflare,[254] Netlify,[255] and Vercel[256] all excel at this task and make it easy to integrate a preview into GitHub pull requests. (FIG 5.12) They best host to-be functionality that is visual, such as a website, deployed sandbox, demo, or component library. If your project applies, a preview deployment can directly solve the unit vs. functional test paradox I just outlined. Automating these ephemeral environments saves you time in pulling down a branch, running the code locally, and verifying the deployed output.

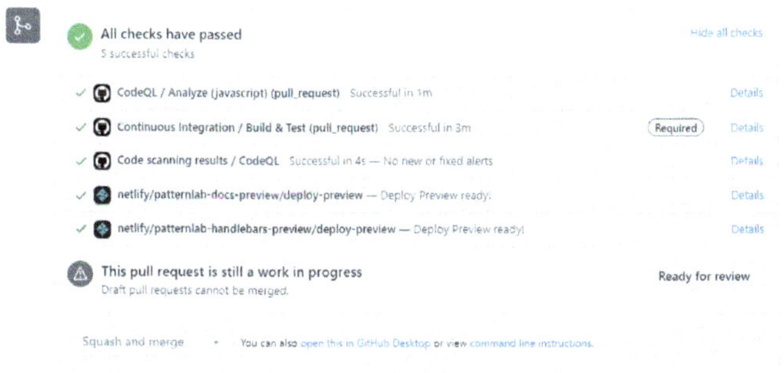

FIG 5.12: A GitHub pull request with two different deployment previews hosted on Netlify to preview project output.

Contributors also see the end-product in a real environment, which might help them contextualize their changes and iterate upon them. Most open source projects opting into this automated preview do not incur a cost, though it depends on your usage and feature set.

Deployment

So far, the value of most of these automations has been creature comforts or tighter feedback loops for contributors. Automated deployment strategies serve a more urgent purpose: to release new code out into the world. It is a mark of high maturity to implement an automated release pipeline. It communicates to end users that you value contribution, and have confidence in your coding standards, review process, and team.

Release early, release often

Release on merge of every change if you can. There might be circumstances that deviate from this pattern, such as a monthly release cadence. Even so, a nightly or beta release might accompany each change. The goal of these releases is rapid distribution and feedback from your community, so this is no time to wait around. It's a very deflating feeling as a new contributor to get your code change approved and merged, only to think "*now what?*" as maintainers sit on it. The `auto` command line tool[257] (mentioned in the Resources section) is built to help release rapidly, incrementing the version of the package according to semver, adding a `CHANGELOG` entry, and publishing all in one package.

There are practical reasons to implement such easy optimization too. Believe me, you want a random pull request to fix a production issue when you are busy and your package is relied on by thousands of other packages. It's never fun (and it could be dangerous) as a contributor to see a merged pull request go undeployed for no other reason than the maintainer needs to manually cut a release.

Generate your artifacts

When automating publication, consider omitting artifacts from Git if there is a compilation step. Checking these changes into Git can lead to confusion during local development. For example, if you are writing your project in TypeScript, you'll likely compile JavaScript build output. Become aware of the process npm uses when publishing a new package. The `files` key[258] in your `package.json` will define what files get bundled into the tar file. Previewing what would be published is done by running the `npm pack` command in your terminal. This creates the whole tar file (installable in its own right), which you'd then need to unzip to inspect. Skip that step by running `npm pack --dry-run`, which will only output to the terminal but not write a file. The `npm link` command can help you test packages pre-deployment too.

What to do when a release goes wrong?

Inevitably, something will go wrong. You'll forget to commit a file. A pull request will overlook something (that should have been tested perhaps). You left an embarrassing `console.log` statement in while debugging. Ask yourself, "*is the release stable?*" The answer to that question may help you determine what to do:

- **Roll back:** revert the most recent change(s) and return the code to a stable state. Depending on the severity, mark the release as unstable by updating any artifacts you can edit easily. Consider deprecation.[259] Change the `latest` dist-tag[260] back to a stable tag.

- **Roll forward:** fix the instability or mistake with a new change. Careful when doing this—you are already under duress.

What you choose is up to you, but try to empathize with your users in this moment. Or recall times that you've consumed an unstable package. How busted is everyone's CI pipeline right now? Is production down? What value exchanges are established, broken, or reinforced by each approach? Maybe it doesn't matter, because you are still resolved to re-establish stability. Keep both tactics in mind as the proverbial shit hits the fan and you'll be more cool-headed about what's right for you and your community.

Maintain Course

Set clear expectations within documentation and tooling. Improve what you need to, only when you need to. Invest in processes that optimize your time. Your footprint should grow with your project's open source activity. An imbalance on either side is stressful—too much too soon feels like over-engineering. Too little too late feels like unmet community needs. Keep your initial value exchanges in mind and revisit them as needed. Allow the work to guide you with intention and eventuality, a boat using the cross-wind to help steer. If you keep at it, before too long, like-minded contributors will find you.

Chapter Six

Open Source in the Workplace

Engaging with open source software is part of your day to day, regardless of your organizational size or mission. To deny its impact and ubiquity would be as foolish as it is false. And yet "96% of the demand-side value is created by only 5% of OSS developers."[261] Where's the disconnect?

Companies aren't paying their fair share.

I suggest further: omitting open source engagement as an integral job responsibility is unacceptable from every perspective.

First, it is shortsighted corporate starvation to not empower their developers to work on open source. They consume this software to further business goals—improving it and their employees helps the bottom line. Second, it is personal closemindedness. Both fail to see the potential of open source as a skill to be trained. A craft to be honed. A *profession* to work purposefully within.

We don't need more open source celebrities. We need more first-time contributors. We need more company tinkering time, corporate open source funds, unstructured learning, and trust to share out what we know. We need more engagement beyond our comfortable spheres of influence. This is the untapped potential energy of our industry. I've seen it ignite, as scarce as a lightning strike. What if we could harness it?

I said responsibility before and I meant it. I see open source engagement as a pursuit worth more than your current circumstances. You build it, consciously or not, as your lasting contribution to the corpus of our field. A field that is still in its infancy. We all gotta eat, of course. But more importantly, *we all gotta survive*. The advancements we make individually and together contribute to the symmathesy of products, industry, and culture. It will outlast your current paycheck, middle-manager, role, and company.

You Don't Need Permission to Do Your Job

Open source software is unavoidable. Contributing to it should be as natural an action as consuming it. Chefs don't ask permission to source new ingredients, tweak existing recipes, or create new dishes—creation is implied in their role. A chef that doesn't do these things is a line cook. Each has their place, but one aligns better with the fluid landscape we operate in as developers.

I remember this mantra, "you don't need permission," came up when component libraries and design systems emerged. Many a thinkpiece was written on whether or not your company needed a design system. Perhaps it was a collective pep talk, or exercise in building toward a crescendo. I built one as a UX team of one, part style guide, part laboratory, without asking my boss. Instead, I *showed* him. He wouldn't have understood why it mattered. And it didn't matter to me, really, if he understood, because the team and product needed this new tool to better organize ourselves. And then, to my observations, the industry stopped debating whether or not we should create design systems. It became the obvious default to adopt and optimize.

We need to do this with open source engagement. We need to take the initiative to better our habitat. There are opportunities everywhere—you are bound to find a corner of the broader community that interests you.

Start small. Instead of buying a maintainer a coffee via a donation, spend a coffee cup's worth of time contributing to their project. Don't ask permission—use that ten minutes as an experiment. Tweak your hypothesis, find your interests on the spectrum of engagement, and report your progress at work. Open source altruism cannot be taught, but it can be modeled. Next time, pair with someone. Show them how approachable it can be.

You don't need permission because your actions already interact outside the four walls of your business, whether you like it or not, into the legal sphere. There are external consequences of your work. It's in your best interest to demand and obtain access to the outside world, as you are not shielded from it:

- You might remember online commentary about the potential liability of Uber engineers developing self-driving cars' autopilot systems when a pedestrian was killed in 2018.[262]
- An individual was indicted by the US Government when his software was used by a third party to commit fraud on the Chicago Mercantile Exchange.[263]
- IT workers in Albania were "placed under house arrest for failing to update the antivirus software on government computers," and could face seven years in prison if convicted.[264]
- Academic work is examining the legal liability software developers expose themselves to under International Humanitarian Law when

autonomous weapons are involved.[265] Did you know International Humanitarian Law is a nicer way to say war crimes?

- A thorough 2014 post-mortem of alleged software deficiencies suggested Toyota vehicles could unintendedly accelerate when the program monitoring the gas pedal crashed. (FIG 6.1)[266]

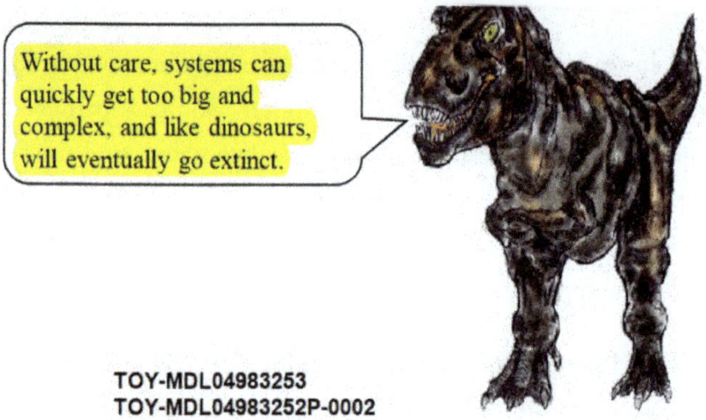

FIG 6.1: An excerpt from an independent post-mortem of Toyota unintended acceleration incidents. This would be funny if dozens of people hadn't died.

Proprietary or open source, your actions as a programmer have consequences beyond keyboard and screen. It's on us to reclaim agency. Agency in saying no to predatory or discriminatory algorithms. Agency in adopting the full spectrum of engagement by default. If your company doesn't let you do that, after insistence, explore your options.

You are part of something larger

The Web as we know it is, what, thirty-some years old? What other consequential human advancement is so close in the rear-view mirror? Do you think engineers thirty years out of the Wright brothers first flight were aware of their historic place at the inflection point of transportation? Did the "Penicillin

Girls"[267] steadfastly harvesting milliliters (weekly!) of the new wonder drug know the weight of their 1930s breakthrough? (FIG 6.2)

FIG 6.2: Two of the six women—Ruth Callow, Claire Inayat, Betty Cooke, Peggy Gardner, Megan Lancaster, and Patricia McKegney—trained to "farm" penicillin from fermenting mold canisters in the 1930s. Industrial scale production wouldn't come for a decade. Copyright © 2017 Women working with penicillin cultures © Museum of the History of Science, University of Oxford

I wonder if there is any other profession with as much richness, diversification, impact, reach, and potential as the Web. We aren't making pamphlets, fansites, and toys here. We aren't selling pants or posting photos. We are engaging in something greater. Do you see it? I don't mean to self-aggrandize here—I am self-actualizing. Connecting the parts to the whole. According to the spectrum of engagement, there is always a vehicle to do this while at work—without having to wade at all into nights and weekends.

I challenge you to adopt this perspective, as Jeremy Keith outlines in his November 2021 talk, *The State of the Web.*[268] We're connecting society, propelling it forward, making it into something unimaginable even a generation ago. Let me emphasize and cast within open source structures. We consume and contribute within our own means, along the spectrum of engagement. But I suggest you should see the work at any particular job as an opportunity to execute your craft at a larger scale, beyond the function, user story, and product of the moment. The open source work you draw from is right there, a relay race with many lanes and runners. The starting blocks and finish lines may feel nondescript, but the baton is always ready for you.

It's easy to throw around oft-guilt-ridden words like *should*. But I aspire to see *should* in a nobler light, if only this one time. You *should* engage with open source because it's a tangible manifestation of something larger and more important. It might sound rote, like, you *should* recycle. But, I argue, respectfully, it's not about you. Keith frames his talk around the 1968 Apollo 8 mission to orbit the moon. He reminds us that astronauts sometimes experience a phenomenon known as the overview effect[269] when seeing the Earth from space. One ceases to see the world's boundaries, substituting this mindset with both a unified vision of the whole and a motivation to think and act in its best interests. We can come close to experiencing the same via the *Earthrise*[270] photo, taken from the same Apollo mission. Here we witness the Earth rising over the horizon of the moon. Quite the change in perspective. (FIG 6.3)

FIG 6.3: Earthrise, taken by Bill Anders aboard Apollo 8 in 1968.

If we choose to see the World Wide Web as more than memes, cat pictures, advertisements, and comment section vitriol, we can grow beyond our individual means and solve truly larger problems. Like crowd-sourcing complicated protein folding calculations,[271] connecting people needing sighted support with volunteers over video,[272] creating access to free wheelchair tech,[273] finishing the knitting projects of late loved ones,[274] or confronting the climate crisis.[275] Open source work permeates these efforts, and so many others, consciously or otherwise.

We are inventing the industry as we speak, for worthy or ill intent—it will be what we make of it. We are so few that individual impact is still felt, in herculean efforts yes, but more so in nudges along the path by us normal folk.

Jenn Schiffer wrote that we are in a unique position to shape and course correct. She wrote, in a now-protected series of social media posts:

> *from my experience over the past decade, workers are making the tech stack decisions more than ever and that's a lot of power i don't think they have processed. and decision-making is labor. don't let the toxicity of "passion" trick you into not considering your job a job. anyway, i prefer community-owned open source too and would love for that to happen without introducing a new flavor of exploitation.*

The intersection of work and open source is no coincidence. Companies have been co-opting open source into their solutions more and more. Some of the original antagonists in the open source movement, such as Microsoft, are now among its strongest supporters. But remember, this is not without looking to their interests too. Employees with hands on keyboard are well-positioned to advocate for and adopt healthy value exchanges. Developers as a form of capital-L Labor are emerging everywhere, from successful unionization efforts at companies like Glitch,[276] Kickstarter, [277]and the New York Times,[278] to software licensing and codes of conduct. The rise of AI while this book was written is challenging nearly every system, nearly any narrative we have. But it's still humans at the wheel (so far), and that means we need self-efficacy. Ethan Marcotte no doubt channels all this better within *You Deserve a Tech Union*.[279]

Everything is open source

Open source is everywhere. Your employer is using it. You are too. If you can conceive of a technical concept, there is probably a project out there looking for help. Open source is everywhere—*like, outta this world, man!*—which I mean literally because NASA's Jet Propulsion Laboratory publishes an open source flight software framework[280] running the Ingenuity helicopter on Mars. (FIG 6.4)[281]

FIG 6.4: A badge that GitHub displays for contributors to projects relating to the Ingenuity mission to Mars.

You don't need to be a rocket scientist to realize there is open source representation up and down our entire stack:

- **Servers:** Apache httpd, nginx, Tomcat
- **Browsers:** Chromium, Firefox, Vivaldi
- **Content Management Systems:** WordPress, Strapi, Drupal, Ghost
- **Static-Site Generators:** Jekyll, Hugo, Eleventy
- **Frameworks:** Next, Nuxt, Svelte, Astro, Fresh, Django, Rocket

- **Libraries:** Express, jQuery, Moment.js, React, css-doodle, Lit, json-webtoken

- **Tools:** AVA, Size Limit, ESLint, Prettier, webpack, Changesets

- **Runtimes:** Node.js, Deno, Bun, Open JDK

The very programming languages we code with every day are often open sourced too. The TC39 committee proposes, documents, and tests new additions to the JavaScript language.[282] Standards bodies like the WHATWG[283] develop the specification for HTML, fetch, URL and more. The W3C[284] contains the CSS working group responsible for the development of new language features.

Take the CSS Nesting Module, at the time of this writing in Editor's Draft status. (FIG 6.5) It's on GitHub to look at and comment on.[285]

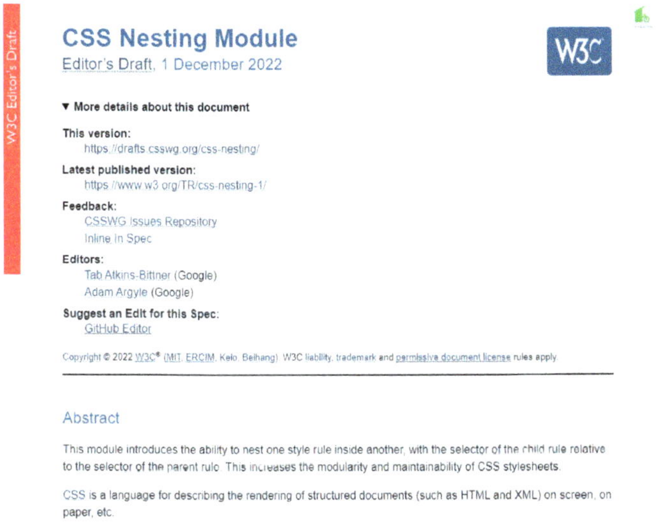

FIG 6.5: The Editor's Draft of the CSS Nesting Module specification open for public comment.

Public feedback on syntax and behavior is encouraged from non-members, allowing us all to participate directly in the standards bodies that govern our foundational tools. You can observe library authors integrating functionality as it evolves,[286] bringing the functionality beyond the theoretical and into more accessible formats for testing. What better way to level up your game than to help make the rules? Impress your coworkers and boss by being on the cutting edge of these emerging features and sharing out their progress toward adoption. The opportunities we have to engage across the spectrum, up and down the stack, create a near-certainty that something will pique your interest. Creating an outlet at work to act on those interests is part of your job. We consume, we contribute. A breathing exercise.

I shared before that this very pathway within the `<picture>` element's journey was what empowered me to realize there was no speed limit on the Web—we are free to devote as much effort as we wish to this shared experiment. My contribution at the time was miniscule in retrospect, but it opened doors I didn't know existed, pushed me to develop skills I never had before. Most importantly, the work connected me with people beyond my local bubble. I see this same spirit of innovation playing out within projects like houdini.how, a site devoted to the education and evangelization of the CSS Paint API. Your work, your team, and your company will benefit from your expanded perspectives and skills.

Gather Momentum

Starting and sustaining an open source effort in the workplace is both an effort in grassroots willpower and corporate organizing. Finding like-minded individuals helps coalesce energy around common problems and shared solutions. Consider forming a guild, unofficial club, meetup group, or even a Slack channel to organize and plan action. Accumulating allies and contacts helps accrue support at higher levels, giving early efforts a longer runway to prove success. We need to be vocal supporters of the efforts, internal and external, that we rely on.

We can do that with our corporate voices and corporate paychecks. Kat Marchán said on social media,

Maintainer after maintainer, the same story: corporate sponsorship of open source work was the main way anyone had time to contribute. So why are we pretending open source is a "pure" community endeavor and the solution isn't just "give them more money from our deep pockets"?

Benefits, or convincing your boss

Contributing to existing open source work is usually an easier task, requiring less permission and more forgiveness, as they say. So when your company chooses to open source its own intellectual property, focusing on the benefits of open source can assuage doubts and concerns. Here's some straightforward arguments you can tailor to your situation. Open source work can:

- **Raise the company profile:** for better or worse, there is a prestige associated with open source work. You are swimming in larger waters. Participating in the marketplace of ideas. Let the analogies run wild.

- **Aid recruiting and retention:** many developers want to be working on open source. Working on projects while working keeps current employees from straying and attracts talent from elsewhere. Potential hires can see what *and how* you work before they even apply. This may filter out candidates before they even reach HR.

- **Skill up employees:** when every action is scrutinized, you have a tendency to improve your game. Engaging on the spectrum of engagement as maintainers teaches you to make more maintainable software.

- **Prioritize the ecosystem:** recognize the altruistic values of open source. This might align well with your company's actions in other sectors, like activism, volunteering, or direct donations to charitable endeavors.

- **Give us "free patches":** just kidding. This shouldn't be in the pitch, though it might happen as a by-product.

Corroborate these ideas with the momentum from your in-house communities, external contributions, and a healthy dose of realism. Remind everyone how much open source you consume. I've found it to be a pretty easy sell when soapboxed a bit as the right thing to do. People want to do the right thing, but you need to show them what that is. Have a plan at the ready. Have a project in mind. Make it their good idea and you've recruited an ally.

Questions needing answers

So, you have their attention. This open source thing might work. Upfront think about common questions. These are the early gotchas that detractors and champions alike may ask about. Being prepared with answers may help you convince a stakeholder of the value of open source, or at least fill any awkward silence with a confident plan that lets you get back to work.

Buy-in questions

Aspects that might stop your efforts before you have a chance to start. The concerns you might hear are:

- Won't this take time away from ...?
- How do we know we are successful?
- Sure. What's next? Where do we start?

Legal questions

Intellectual property and legal hurdles may need resolution before open sourcing anything you do on company time:

- Do we need a CLA process? What message does that send to the community?

- How are trade-secrets, company data, or patents handled?
- What license do you want to use?

Contribution-based questions

Nitty-gritty details once you are up and running:

- Do we plan to only contribute new work? How do we enable employees to contribute outside the organization? When do we fork an existing project?
- Which projects make the best open source contributions? Are there some that should be avoided?
- What do we do when people leave the company?
- What if we don't like the contributions we get?
- How do we ensure quality? (apply some of the automation lessons from earlier)

Security questions

Safely engaging in open source communities:

- How can we make contributions and releases secure?
- Do we need two factor authentication on developer accounts?
- Should automated processes use developer secrets or single-purpose accounts?

Having answers to these questions will serve as a good first draft of real company policy. Your answers don't even have to be right, yet, that's for the whole team of stakeholders to align on. But it does demonstrate a firm grasp on the breadth of issues your company will encounter beyond `git push`.

Slow and steady

Go slow at the outset to go further in the long run. This is a big flywheel to get in motion. Your company's open source presence will grow with intentional effort and community-building. Sustain contributions once launched, lest your open sourcing efforts become a graveyard. One of the best ways to do this is to continue to use the software yourself, sometimes called unappetizingly "eating your own dogfood."

I'm excited about the prospect of open sourcing code I write at work, and am happy for the opportunities that present themselves. But I acknowledge that not all projects are right to open source. Don't pursue open sourcing a new project just for the sake of doing it or to check a box. Don't open source something because your manager told you to (or because I did). In both cases you'll be chasing a feeling of accomplishment that needs to come from elsewhere. Yes, we are trying to get buy-on from management, but not at the expense of readiness and resolve at the engineering level. Your team needs to be mature enough to operate at the maintainer level, transitioning many processes to an outward, public perspective. Open sourcing at the outset of a project may be easier than adapting an existing project. Have these conversations with your team and stakeholders. Know where everyone stands so you can direct the conversation toward the best outcome for you, your team, company, software, and community.

I've found success, when the stars aren't aligned, open sourcing what I'd call the orbital bodies. Trying the whole planet, the flagship product, might be more than you can handle, or it might be too integral to the business model of the company. But an internal tool, generic libraries, an automation, these things are interesting candidates. Spotify open sourced their Backstage developer portal[287] years ago and recently released some premium plugins. Many other companies have vast catalogs of projects that seem ancillary to their core customer offerings, and that's okay! You can do that too.

Innersource

Still not ready? That's okay—there's another coding strategy your company can adopt to build community. *Innersource* is the conscious application of open source principles to projects in the workplace—though they may not be wholly open sourced themselves. This smaller scope and population size is more manageable. Y'all should already be friendly, familiarized, and aligned to shared purpose. Innersource efforts still need good fundamentals like documentation and expectations exchanges. It's a great outlet to practice and hone skills that benefit everyone, open sourcing or not. A spark of altruism and sponsorship from management helps get innersource projects off the ground.

But developers and managers need to know that successful innersource projects don't happen by accident either. This isn't the absence of all governance—this is intentional stewardship. Innersource projects are both products and communities. That means you might spend development time working on tasks that *only* improve community, like celebrating contributions, conducting webinars, open office hours, etc.

The shared context so important to newcomers in the open source community is important here too. Don't let information become siloed or hidden away behind private team areas. Innersource projects need to be open by design, not by default. These projects need high-engagement to sustain themselves. That's why it's important to pick the right projects to rally around. They need to solve a generalized problem within your company and have enough users and applications to see continual development. A component library is a great example.

Maintainers need to plan avenues for contribution along the spectrum of engagement. Internal contributors need headroom to spend a bit more time on a story so that they solve for everyone and not only their product. Management needs to understand that resource allocation, cross-team goalsetting, and prioritization is harder the more teams you add to the equation here. A centralized group empowered to govern and shepherd contributions is important. You need a brain to keep it moving in the right direction, a janitorial staff to keep the lights on, and a steady roster of motivated helpers to keep development rolling.

The InnerSource Commons organization[288] brings together companies practicing innersource development to share and advocate for successful patterns. They've documented over twenty patterns that organizations can adapt to their needs, covering four concept areas: Begin, Adopt, Grow, and Scale.[289] If this looks reminiscent to the spectrum of engagement by now, I'll consider myself successful. I've implemented and observed the effect these patterns can have on the health of a project. It's really heartening to see that community can form around process, not just tooling.

An innersource project can be a gateway to larger open source efforts, as your team will learn how to manage and communicate change for a wider audience. If it happens, it happens. You'll know when you are ready.

Open Doors and Minds

I'm not trying to push you and your team to engage with open source. You need to find motivation within yourselves to do that. But don't convince yourself that you aren't affected by the outside world—and in acknowledging this, take action to shape it toward a more just, equitable, and useful place to work. We draw from the open source community every day, and I don't mean owls this time. We are building something that will outlast all of us—it's hard to even fathom. Recruit allies to advance your interests, and look beyond your current role to the larger currents of work flowing around you. Focus on the little things at first, what you can manage. The community will be better when you show up, even if you start small.

Chapter Seven

Open Source Sustainability

You've started this journey, made measurable progress from the trailhead. It feels good. But also the sweat of exertion reminds you that its work to get where you've gone. A day-hike is as good as a lunch walk is as good as a guided expedition. But the view from wherever the trail leads you reveals further mountains to climb, as far as the eye can see. This evokes the Haitian Creole proverb, "dèyè mòn gen mòn" meaning "behind mountains, there are mountains." Challenge, and opportunity. If you let your imagination run wild, you might wonder or worry, *"now what? What's next? When do I stop?"*

Luckily, open source contribution isn't the Seven Summits[290] of mountain climbing. There is no checklist to compel you toward someone else's idea of success. I've suggested that planting your flag at the summit of an open source mountain is not the end-all goal—value comes from the experience itself and the trails we blaze for the next travelers. So our mission should be yes, to keep our destination in mind, but not without a conscious understanding of the preparation, planning, and surefooted steps we can take to get there. No one sprints up mountains, and no one can run to open source success. We need to act within patterns that we can sustain. We should work with a mindfulness of teamwork, shared purpose, and acknowledged risk. And we must accept that someday we will choose to stop.

Dispassion by Design

Open source projects don't need to evoke thoughts of late-night hacking sessions, crowded coffee-shops, or blockbuster developer relations parties. These scenes convey to some the "passion" for coding. It's rubbish. Programming can be a job. There is no altar at which we must sacrifice to prove our worth. It can be as Josh Clark puts it "boring,"[291] predictable, and well, work. He suggests design system work "is not to invent, but to curate." Open source can align with that, while still holding true to our larger concept of open source as a movement, a force for change, a new medium we get to shape in mind and body. I hope I've equipped you with enough tactics and perspectives to feel comfortable in this position.

You'll be confronted with plenty of noisy examples to the contrary, however. *Someone open-sourcing harder than you.* I laughed to myself writing that. You won't be making or contributing to any software if you burn out. The concerns of a project will tug at you, fueling doubt and imposter syndrome. It can feel like and become a real burden if you let it. Let's call attention to some of these. In the end, its purposefulness, not passion, that will sustain you.

You won't make everyone happy

Inevitably someone will be dissatisfied with your work, by choice or omission. Establish clear expectations so users know the territory they are stepping into. The gradation of relationship you design is up to you. I suggest that the community you create is a responsibility. In connecting we further everyone's goals. Others see free code as that and nothing further. There is no relationship. There is no "open source supply chain."[292]

How people interact with your project is a consequence of the expectations you set. Happiness shouldn't necessarily be the measure of success—but clarity in value and telescoped outcomes. If you release breaking changes, are unresponsive, and leave the code to stagnate, consumers will not come to expect quality. Careful improvement and intentional communication create more confidence. Work at what patterns you want to establish and reinforce with your actions. Past success is an indicator of future performance here. I'd bet on those odds.

Your API won't fit every use case

The reasons you create your project are likely clear at the outset. But you won't cover an exhaustive array of use cases. You'll forget something, or the environment will change around you. Someone will want slightly different data flow, output, or file format. They might even be willing to contribute it. Or more likely, you know exactly where to introduce that `if-then` check.

These changes are tempting to make, why not accept them? Afterall, I've praised the virtue of the improv artist throughout this book. Accepting of change and iteration is a welcoming force within an open source project. But you need to think through the consequences of each change introduced into the project. Luke Wroblewski quips "every mobile app attempts to expand until it includes chat. Those applications which do not are replaced by ones which can."[293] Expanding the core offering of your project can ironically make it less desirable. I've been in talks with teams that turn down software because it does too much, introducing single points of failure. The ability to compose small packages together is the real power of healthy tech stacks.

The command line arguments, config file options, and modes of a project add up to cognitive overload for newcomers and maintainers. Divergent paths create multiplicative paths that pressure you to test every permutation in perpetuity. Make sure you accept changes for the right reasons. Don't compromise the balance of effort you've cultivated within your project. A change too far may diminish your mental model of the software. An innocuous pull request might omit documentation, or it might introduce new flows you hadn't had to consider.

Projects need time to acclimate to foreign code. If you can, make sure substantial contributions come with some upfront guarantees that the contributor will stick around to explain and help. The Innersource Commons defines a "thirty day warranty" pattern[294] for just this purpose. It's easier to pull off within an organization, but applying the same concept to an open source project is fair game. If nothing else, it communicates to contributors that you value their work but need their help in supporting it. Hopefully they'll be flattered, and since they are already engaging with a large effort, you know they are motivated to see the project succeed.

My time with Pattern Lab included plenty of moments where I wasn't mature enough to accept use cases beyond my mental model. I had a mixed bag of success in accepting change. The API made sense to me as-is, why change it? I was lacking in empathy and imagination. Some changes I accepted in pursuit of improvement, but the coding style was terse and unfamiliar. It made it harder for me to feel ownership. The contributor thankfully added tests and stuck around for quite some time, but nonetheless it always felt "not invented here."[295] I should have identified my limits, confronted these feelings, and communicated my need for help long before I hit the merge button.

Beware assigning too much of your self-worth

This is what I'd assign as the most direct effect on my own burnout. It cannot be said enough: *you are not your code*. Problem is, with open source software, more often than not, you don't have a manager or set of colleagues to remind you of this—it's only you and the backlog. I hope that you can look back on the code you wrote yesterday, last month, or five years ago and laugh, not feel embarrassed. Give yourself permission to be wrong and have learned. You know more now than yesterday and you'll know more tomorrow than you do today. Alice Goldfuss encourages us to take note of moments when we can look back and see self-improvement, or observe when the novel pattern becomes muscle-memory.[296]

Every issue cannot be an existential crisis or you run the risk of becoming one too. Walk away when you are done for the day. Afford yourself thinking time, yes, but not at the expense of everything else. Walk your dog. Play with your kids. Eat some pizza. Letting your psyche rise and fall on the fortunes of your open source work will surely be a wager you don't want to make—the house always wins.

You'll be exposed to stressors. I try to recall that:

- People will copy you; accept the flattery
- People will steal your ideas and remix it; consider it product feedback
- People will nit on your code; open your mind to improvement
- People will want things faster; encourage them to contribute

Emotions will come from every interaction you have with your community. Acknowledge and make space for your reaction, and then respond when its healthy to do so, if at all.

Walk to emergencies

I trained briefly to be an EMT Paramedic. I still remember my instructor telling us never to run when on the scene. The time savings doesn't really matter in most circumstances, and you'll arrive at the patient with elevated vitals. Not the best way to start intubation where one slip can break teeth. An actual EMT Ross Cohen expanded on this to also mention[297]:

1. *Running is risky. If we trip, fall, get hurt—now there's two patients.*
2. *Running to a patient prevents us from spotting hazards on the way*

in. Remember, we're walking into an unknown. We must carefully observe and assess for danger. On the walk in we'll notice the downed power lines, the room full of people passed out from a gas leak, the dog protecting its injured owner, the hoarder's junk on the floor when we round the corner, the attacker who assaulted the victim, the weapon lying next to the bystander, etc.

3. *Running makes it harder to keep our cool and operate at peak effectiveness. We need to be the calmest person in the room. Everyone takes their cues from us. We need to think clearly and act deliberately, decisively, expeditiously, smoothly.*

Our open source work needs to be likewise methodical. If everything elicits a fight-or-flight response we would exhaust ourselves and desensitize our notion of real urgency and danger. If an issue comes in, on someone's dev environment, and there's a workaround, perhaps we can take a breath and think through how we got here and how best to act. The triage skills you hone over time will shield you from too much worry—a negotiation of user empathy and your own safety.

Rarely in open source is urgency a real problem. It might be *very noisy* when CI systems break across the world, but it's more nuisance than crisis. Libraries that are consumed need to be installed as upgrades by customers, pass through any number of dev and QA environments, and be deployed to their production environment in order to cause real harm. So allow yourself to get comfortable before you debug. Asking "what's changed recently" is always a good start. Understand the impact of the problem as a step towards building remediation options. If there's true instability, revert and roll back to a stable state if you can, saving the post-mortem "why" for after service is restored. If there's an omission or unexpected behavior, gather more information, maybe write a failing test first.

Contributors will come and go

The spectrum of engagement is a highway with many lanes, onramps, and offramps. Not all roads lead to the same destination, and not all drivers have a shared journey in mind. Efforts to onboard and retain contributors should be built-in byproducts of your investment in good documentation and governance. Remember, we want a new contributor to be just as advantaged as the first few. What a contributor needs to know should be discoverable, in writing, and up to date. Investing in self-service context-building will always pay off, because it scales better.

I know because I can remember the enthusiastic messages exchanged between me and new contributors, telling them the same things I told the last one. It felt good in the moment to be making new connections, no doubt. But more often than not, those personalized training sessions didn't shape lasting contributors. This can be deeply disappointing if you look to these newcomers as a tag-team partner too eagerly. Respect and honor the time others can give you, meet people where they are, and open as many lanes as you can.

"Sustaining Maintaining"

Dave Rupert gets top billing on this section for his inquisitive blog post of the same name.[298] He outlines the challenge of burnout and scale we've covered here and suggests how to chip away at these problems. I appreciate his earnest sharing of personal experiences too. Of particular resonance to me is the acknowledgment that when burnout encroaches, we seek refuge elsewhere to cope. It becomes a stall that's hard to pull out of. Some of the things we open source grow up to be more than side-projects, toys, learnings, or errant ventures. It's how people make a living. It's a responsibility. It's a burden. It's a relay race we must acknowledge we can only run so many legs of.

I agree with Dave that money and time yield the best results. I posit that corporate engineers have the largest role to play here, whether they know it or not.

He says, "Open source maintenance needs a manual," and in my wildest dreams this book could meet that measure. If you've open sourced something and are struggling, find comfort in these pages and the resources they point to.

Many others are committed to improving open source sustainability too. The Sustain project[299] was conceived in 2017 as a sort of "unconference"—meetups and loose working groups focused on advancing topics like governance, documentation, marketing, licensing, ethics, and more. They also host several podcasts[300] worth checking out across subject areas. The Community Health Analytics in Open Source Software (CHAOSS) project[301] is a Linux Foundation project "focused on creating metrics, metrics models, and software to better understand open source community health on a global scale."

I am energized by the increased interest and awareness in sustainability. I think we as a community are now understanding that most software problems are people problems, and that us people are worth supporting, patching, and funding.

Pace and piecework progress

When I'm not coding or corralling kids, I pretend I'm a carpenter. During the pandemic I insisted I needed to hand plane[302] my workbench and the desktops I made for me and my son. I found two-thirds success so I guess I cannot go pro yet.

I read it as a sign of encouragement, though, when my wife bought me *The Why and How of Woodworking*, by Michael Pekovich. What struck me was that the ethos of the early chapters could be written as a manifesto for healthy open source development. Look at what Pekovich calls his rules, from the chapter entitled, "Make Shop Time Matter"[303]:

1. *It's not the number of hours, but how you use them*

2. *Make it a habit*

3. *Do your thinking away from the shop*

4. *Make sure it's worth building*

5. *Why the little things are the "big things"*

6. *Shavings or sawdust?*

7. *Our mindset makes a difference*

8. *A new technique and a new tool with every project*

9. *Rethink perfect*

10. *Don't rush, but don't stand still*

11. *Respect the miter*

12. *Sometimes you just need to let it go*

Sawdust and miters might be a stretch, though I could convince you we are talking about dependencies. I'd love to dwell here longer, but I bet you can see it too. Open source work, woodworking, needlework, footwork, teamwork—all work—requires time and craftsmanship. We need to treat every task as important, additive, and interconnected to the whole. Cutting a corner can come back to haunt you. Blunt tools or outdated dependencies slow you down and can be dangerous. The shop or Git branch left untidy between jobs will sap your productivity the next time you find a moment. Or worse, contributors will find your mess first.

The pace of your contribution does not need ridiculous velocity. It can be modest, mechanical, prioritized, and steadfast. Dispassionate care toward your project and the open source movement will give you staying power and more control over the arc of work you engage in.

Dog-fooding

The paramount means to sustain development of an open source project is to keep using it. I won't pretend this is profound. But it's something I realized too late with Pattern Lab. I drifted away from daily driving my own software, and in doing so slowly became insular. It wasn't intentional—but my career took me away from it and the spare time I could devote to it wasn't enough. I *thought* I knew what the concerns of the community were. And even so, I couldn't muster the sustained energy to make progress.

Acknowledging the stagnated development, I wrote "I, Maintainer." It was a short admission, admonishment, and call to act in accordance with community interests. Fashioning myself after Isaac Asimov, I coined, "Muenzenmeyer's Laws of Open Source Maintenance." I'd like to revisit them, five plus years later[304]:

1. *A maintainer may not harm a user or, through inaction, allow users to come to harm.*
2. *A maintainer must shepherd their project and empower the community to do the same.*
3. *A maintainer must protect their own existence as long as such protection does not conflict with the First or Second Laws.*

By the way, simply writing them didn't change the trajectory of my burnout. Not long after, I retired from active development of Pattern Lab. And these "laws?" How do they hold up, even to my own scrutiny? Well, the First and Second law still feel unassailable to me. I don't buy into to the idea that open source is a dumpster that you rummage through for treasure. I've made myself clear that it's more than that. The community you create, intentional or otherwise, is inextricably linked to the act of sharing it in the first place. Impact over intent.

The Third law, however, I got wrong at the time. It reads as a doubling down that community is more important than self, when that is flat out backwards. The rest of the post even walks back the law right away:

> *When a maintainer chooses their reality over the reality of their users, they betray the trust put in them by users. They are prioritizing their own needs.*
> *Good. No one else will as easily.*
> *But to do so without implementing a healthy, sustainable community is acting against the interests of the project. It is an abdication of the very responsibility they have earned.*

I guess you could spin "protection" as a graceful exit, but I think I'd rather be obvious:

> 3. A maintainer must protect their own existence ~~as long as~~ *even if* such protection conflicts with the First or Second Laws.

I feel that my experience shaped the theory that corporate developer contribution is key to long-term success. I also had to experience the failure of unmet expectations as a maintainer. Engagement must come from a healthy community of users, and you must count yourself among them. You build empathy with your users by experiencing the same constraints, triumphs, and shortcomings as they do. This broadest realization of the spectrum of engagement has us acting as consumer and contributor from moment to moment. You need to be connected to the concerns of your community.

Funding

"Money is the least-effective language in the open source sustainability conversation. But it's a language we use a lot, because it's so widely understood," says Duane O'Brien, former director of open source at Indeed.[305] As our project grows and more time is needed to care for it, it feels natural to want to be compensated. No one should question you for monetizing your skill and stewardship.

Accepting funding is a direct solicitation for support. You're creating a pathway for others to empower you. What promises are you making with donations? How will you disburse them to collaborators? Think through if money is the real bottom line for you. Would you get more piece of mind from more contributors instead? The same platforms mentioned earlier are good places to start. Open Collective, Patreon, GitHub Sponsors. More will certainly sprout up. GitHub launched an accelerator program[306] in 2022 that pairs a stipend with access to industry experts to empower open source maintainers to focus full time on projects. The company Clerk is sponsoring full-time development of the next version of the Colin McDonnell's Zod validation library as a temporary "OSS Fellowship." [307] I hope to see more novel funding efforts from corporate interests.

Legacy, sunset, and succession

We wouldn't be responsible maintainers if we didn't know when to part ways with a project. We might do more harm when we linger or stagnate. Do you do your best work when you no longer have the time or heart for it? If we have the resolve to rip the Band-Aid off, the community will be better for it.

How do you make the decision to step away? I joke that parenting is fifty-one percent worth it. It's that one percent that makes all the difference. *Kids, it's more complicated than that, but boy do we have our moments.* Open source can be a relationship not unlike parenting at times. When something's *your baby*, your mind can make a pretzel of any rough situation to justify sticking it out. You go through the motions, but don't find fulfillment in it. A lot of feelings can get tangled into the work. It might seem endless and inescapable. But this isn't raising a child or caring for a loved one—those responsibilities are sacrosanct. Open source should have off ramps rooted in personal safety as well as normalcy. So try to think about the specific tasks. If the thought of sprucing up the issue log, helping users, cutting releases, documenting everything, and (I suppose) coding doesn't consistently bring you fifty-one percent contentment, you might want to consider a hiatus. Maybe that threshold percentage is greater for you, you with the high-expectations. And they should be high, because we all have limited time and don't intend to squander it.

If you decide to walk away and there is no one ready to take the baton, make it clear within your project that you are looking for a maintainer or that the project has run its course. Coalesce what community you have around a fork perhaps. Most code repositories now have a means to archive a repo as read-only. This communicates clearly that the project is not interested in future engagement.

Hopefully your project has attracted a community willing to take up the mantle of responsibility. Their enthusiasm and readiness to succeed you is the final artifact of the training efforts your actions influenced. Be clear about what engagement expectations you have as an emeritus member. Take comfort in knowing that while you may no longer be a maintainer, the spectrum of engagement remains open to you.

Where We Go Next

I asked Abby Cabunoc Mayes if she'd speak with me as part of writing this book. She's been program lead of Mozilla's Open Leaders program (where I was fortunate enough to join a cohort in 2018). She's spoken at conferences including OSCON and StrangeLoop, keynoting EuroSciPy and Community Leadership Summit. She organizes with the folks at Sustain and is an OpenJS Foundation Board Director. Abby is also an Open Source Program Manager at GitHub. To be clear—Abby has unparalleled awareness of and advocacy for open source communities.

Abby was forthright with sharing that we as an industry are still figuring out sustainability. "Open source is so new. We are at an inflection point. We're so reliant on open source right now, but the models don't seem to be working. I think there's a model of open source that we haven't tapped yet." The transactional model of open source contribution feels just that—a buying and selling of value. It's business. It's impersonal. *No release in 4 months? What have you done for me lately?* We need to manifest a purposeful, larger perspective, like Jeremy Keith inspired with his *Earthrise* analogy. Abby challenges us to think of open source as a movement, not pay to play. Don't focus too much on the venture-capital-funded heavyweights.

Abby highlighted a Twitter thread that suggested that major tech companies were not structuring job descriptions and promotion guidance with open source in mind.[308] Maintenance doesn't seem to be a strong selling point to leadership. It's hard to quantify how a bug fix impacts the bottom line (until the lack thereof does). Min-maxing to get ahead makes sense in our society, but it's not without cost to the commons. This threatens the very idea that we can aspire to be more than the needs of our current selves. We need to demand recognition from employers that open source engagement is a job function—one that's worth hiring, training, retaining, and rewarding.

She also reminded me that maintainers are leaving the ecosystem. This could certainly be burnout or natural attrition, but increasingly in our thirty year old Web, age will become a factor. The maintainer of the `minimist` CLI tool abruptly left the community and outright deleted their GitHub account and repositories. This package has twenty-one thousand direct dependents and almost two hundred million monthly downloads.[309] It cannot *just* go away. The community stepped in, as it usually does, to steward this and many other packages. But we're missing something here. We cannot discard this work like orbital debris and act surprised when catastrophe occurs. We need an offramp for these maintainers that is safe and managed and has healthy value exchanges. A retirement where they don't fear their inbox.

The 2024 xz utils vulnerability[310] was enacted in large part due to socially engineered pressure to add another (malicious) maintainer. I myself have been given push access to projects after a few months of work. Maintainers are starved for help. Those fault lines of inclusion and exhaustion and exploitation and value: they are building up pressure. Something's gotta give—and it should be all of us, and companies, giving a damn.

Abby continued, "we need to make room for casual contribution. With time-boxed scopes. More obvious onramps and offramps. We are connecting and getting places faster than ever, but haven't strengthened the muscles of working together as a group." All this echoes the spectrum of engagement, and the rapid pace with which our social coding has exposed fault lines always present in any community. "It needs to evolve," Abby says of open source, "but in evolving what open source means, we need to think about what parts we are leaving and what parts we bring with us." She cited how the push for codes of conduct led to the pressure to truly enforce them, which revealed the need for better overall governance. Advancing expectations. She sees the opportunity amidst the challenge. I hear weariness at times, but there's a certain optimism too. A deep optimism that you can reliably draw from once you've seen open source in its best lights, connecting with strangers and advancing common cause. It's a strong feeling to chase.

I asked Abby what motivates her to contribute. She told me of her grandmother Paz, who passed away from cancer. Through a series of events, Abby found herself writing software to advance and curate cancer research. "Open source was not the norm. Researchers would fudge their data or omit data so they could publish faster. It was hindering innovation." People at academic conferences would give presentations based on work done in Abby's software. "'I use that every day', almost everyone said that," according to Abby, in those early days of her career.

She told me, "I think that's what's best for the world."

I mean, really, what else is there to say?

Conclusion

In too many team meetings or dinner table conversations, I tend to wax poetic about a higher concept than present company cares to hear about. I'm sure there's a word for it. It might be hard to endure, and for that I apologize. But here goes—I have you right where I want you by now.

Open source is the paradigm-changing, self-sustaining, boundary-pushing paradigm of the World Wide Web. How lucky we are! We are fortunate to be alive during its inception, adoption, and early evolution. We see its promise and its threats. We enjoy its delights and lament its shortcomings. More fortunate are we to influence this evolution and its impact on the world around us. Michele Barker describes this relationship in her *Humane Web Manifesto*[311]:

> *we should all have the opportunity to create, publish, build and consume content on the web. Like a functioning society, we take what we need, and we contribute what we can. We are citizens of the web... We* ***all*** *shape the web of today, whether or not by choice. Just as opting out of society is virtually impossible, in the world of today we cannot help but be Netizens. Therefore we all have a duty of care and respect, and we all have a stake in a humane web.*

We can be participatory in the greatest symmathesy—mutual learning through interaction—humans ever built. The Web, open source, and our relationship to them, are at once part calling, part celebration, part crucible, part caricature. It's all there, if only you *sonder* a glance.

We don't occupy more than one of these spaces at a time. We can move fluidly along the spectrum of engagement, project to project, moment to moment, mood to mood. Opportunities to engage will present themselves through the constraints of our work, the proximity of our needs, and our willingness to share. With perhaps a bit of willpower or altruism, you can push through the urgency of your day and do that one extra open source thing.

Start there. Start with the smallest thing. In the end, it's all small stuff. Each forest a tree. Each tree a branch. Each branch a leaf. The winds of change blow and bend and break, but from these challenges we yield stronger roots and wider trunks.[312] Walk outside. Find a path. Patron a community space. You can create your own or find a welcoming one already established, in need of your time and care. Our cultivation, even in such controlled conditions, is connection and chaos and movement. Movement forward. Duane O'Brien said of open source, "if you want the forest to be there next year, you have to care for it this year." Wherever and however you engage with nature, be mindful of everyone else relying on it too.

Resources

One thing's for certain, with open source software, there's always another link to click. I hope these pages prepared you to dive in along the spectrum of engagement. If you need a little inspiration, check out these additional destinations.

More info, updates, resources, and errata at approachableopensource.com

Open source

Read how the Netscape Navigator browser was open sourced, before the term even existed. Collaboratively written by Jim Hamerly and Tom Paquin with Susan Walton within *Open Sources: Voices from the Open Source Revolution.* (https://www.oreilly.com/openbook/opensources/book/netrev.html)

Read Christine Peterson's full account of coining the term *open source.* (https://opensource.com/article/18/2/coining-term-open-source-software)

Watch Jacob Thornton ad-lib the history of the open source software and how he got caught up in the wrong value exchanges during his own career. Stay for the cute puppy analogy. (https://www.youtube.com/watch?v=UIDb6VBO9os)

Matt Rickard defines some of the open source postures businesses adopt and their motivations. (https://matt-rickard.com/short-taxonomy-of-open-source-strategies)

Domenic Denicola outlines the problems with the annual Hacktoberfest event wherein the value exchanges are incongruent between maintainer and contributor, leading to low-quality pull request spam. (https://blog.domenic.me/hacktoberfest/)

See class-act governance from the Moment.js team in outlining reasons *not* to use their project anymore. (https://momentjs.com/docs/#/-project-status/)

View all the free and open source licenses as tracked by the Linux Foundation, along with their status with the Free Software Foundation or Open Source Initiative. (https://spdx.org/licenses/)

Read the essays Mike Pennisi thoughtfully articulated about the seven year long relicensing peril the JSHint project underwent. The balance of empathy, introspection, and acumen is what I hope to see more of in the communities I engage in. (https://mikepennisi.com/blog/2020/you-may-finally-use-jshint-for-evil/)

Check out the FOSS Funders, a "group of companies working together to make it weird not to fund Free and Open Source Software (FOSS)." (https://fossfunders.com/)

Learn about Contributor License Agreements and Developer Certificates of Origin, and how they differ from one another. You might choose to use one in your own projects, and before long will encounter them in the wild too. (https://opensource.com/article/18/3/cla-vs-dco-whats-difference)

Producing Open Source Software by Karl Fogel is a fantastic read and about three and a half times longer than *Approachable Open Source*. If I hadn't had a target wordcount in mind at the outset, I could see any writing on this subject sprawling towards those lengths. You're sure to find great insight there, and I love the emphasis on accessible translations and formats. (https://producingoss.com/)

Community

Perspectives on the effectiveness of protest via "inconveniencing" the public. Brandon Terry put it best within the article: """I think the best protests are the ones that teach us about the injustices that they're trying to attack." (https://www.wgbh.org/news/post/fine-art-inconveniencing-public)

Sameera Kapila writes extensively on broadening community within *Inclusive Design Communities* (https://abookapart.com/products/inclusive-design-communities).

Inclusive 101 by Microsoft Design is a go-to primer on the subject of accessibility, and how inclusive design benefits everyone. I've taken to heart the situational impairments they point out, like high-contrast outdoor environments or one-handed new parents juggling a baby. (https://download.microsoft.com/download/b/0/d/b0d4bf87-09ce-4417-8f28-d60703d672ed/inclusive_toolkit_manual_final.pdf, PDF)

Barak Obama reminds us of *citizens*, "ordinary men and women, determined to forge their own future—who throughout history have sparked all the great change and progress." (https://obamawhitehouse.archives.gov/blog/2014/09/23/most-important-title-citizen-president-obama-importance-civil-society)

Michelle Barker's *Humane Web Manifesto* builds on the citizen comment, reminding us of n*etizens.* There isn't a great term for open source citizens, but I'll leave this here. (https://humanewebmanifesto.com/) and (https://en.wikipedia.org/wiki/Netizen)

Technology

Pattern Lab by Brad Frost, Dave Olsen, and open source contributors helps you design, create, test, document, and distribute component-based patterns. It was storybook before it existed. (https://patternlab.io/)

Atomic Design by Brad Frost is a design and development methodology that encourages web developers to compose hierarchies of larger and larger components. (https://atomicdesign.bradfrost.com/)

The CLI and package registry known as npm is not in fact and acronym for Node.js package manager. (https://github.com/npm/cli#is-npm-an-acronym-for-node-package-manager)

Read more about the history of the World Wide Web and open standards on MDN's brief primer on the subject. Of particular interest is the "don't break the web" section as it applies to old projects. (https://developer.mozilla.org/en-US/docs/Learn/Getting_started_with_the_web/The_web_and_web_standards#dont_break_the_web)

Check out this xkcd comic that calculates the return on investment for automating a task, based on the time saved and how often you need to do it. Some things just aren't worth automating if it takes you too long to implement. (https://xkcd.com/1205/)

Read the W3C's Web Platform Design Principles, especially the *Priority of Constituencies.* (https://www.w3.org/TR/design-principles/#priority-of-constituencies)

Managing dependencies is greatly aided by tools designed to integrate into your workflow. I recommend:

- **ncu**: a command line interface, like the built in `npm outdated` command, but with more configurability (https://www.npmjs.com/package/npm-check-updates)
- **Dependabot**: GitHub's integrated tool to create pull requests for new versions of dependencies and vulnerabilities (https://docs.github.com/en/code-security/supply-chain-security/understanding-your-software-supply-chain/about-supply-chain-security#what-is-dependabot)

- **Snyk**: similar to dependabot, but independent of GitHub (https://snyk.io/)

I've used and seen these website options work well for open source documentation sites (omitting the skinned GitHub pages approach, which only renders a project's README):

- **Eleventy**: static site generator with dozens of deployable templates, many plugins, no heavy-handed conventions, and an A+ community (https://www.11ty.dev/)
- **Nextra**: Markdown and React-based static site with built-in navigation, dark mode, code highlighting, internationalization, and simple search (https://nextra.vercel.app/). I like this is as a sweet-spot. Kinda heavy under the hood though.
- **Docusaurus:** an advanced site builder, with big differentiators stemming from documentation versioning and enhanced customization (https://docusaurus.io/)
- I omit Jekyll and Hugo here only because they don't have the best Windows support last I looked.

Documenting change is aided by tooling that can guide our semver-compliance, but all require setup and adherence to reach their fullest potential. Consider these tools when you want to level up your commit-ment:

- **Commitizen**: a CLI, if you insist on documenting change at the commit level (https://github.com/commitizen/cz-cli)
- **auto**: use pull requests to set semver and create `CHANGELOG` entries. Extensive plugin framework. GitHub only. (https://github.com/intuit/auto)
- **Changesets**: use a structured markdown file with each pull request or merge request. Refined form of the concept introduced with auto, as

release notes are now reviewable too. (https://github.com/changeset s/changesets)

Want to dive deeper into Git? Check out *Git for Humans* by David Demaree. Like all ABA books, this is one resource you can commit to without much conflict. (https://abookapart.com/products/git-for-humans)

We talk testing a lot within the book—but can't quite devote direct attention to it. This primer from BrowserStack on the types of testing helps differentiate when to use each approach. (https://www.browserstack.com/guide/types-of -testing)

The Testing Library project aims bridge the gap between unit and functional test suites, with support for languages and libraries like React, Vue, Angular. Their API mimics user behavior, and discourages any introspection of internals. Their motto is "the more your tests resemble the way your software is used, the more confidence they can give you." (https://testing-library.com/docs/)

Random

Sonder feels a lot like the unofficial etymology of the déjà vu-like Baader-Meinhof phenomenon. (https://www.dictionary.com/e/tech-science/baader-mein hof-phenomenon/)

I really dig the approach Maggie Appleton takes to her writing, where ideas are nurtured from seedling to mature evergreen thoughts. This is known as *digital gardening*, and involves formulating ideas in the open instead of the big ol' blog post reveal against the social-media firehose. Maggie also writes a history of digital gardening, which is a fantastic read. (https://maggieappleton.com/g arden-history)

Acknowledgements

In primary position is A Book Apart, an unmistakable influence in shaping mine and many others' world-view on web design and development. To Jeffrey Zeldman, who's inspired, had random conference lunch with, and now entrusted me to add a new chapter to the spectrum—thank you. Unfortunate that mine and ABA's partnership couldn't bring the book into its final form, but they were with me in those critical first moments.

To Lisa Maria Marquis for the direct, sustained, and thoughtful feedback throughout. I lost count of moments where you pushed me and manuscript to be better. To Mat Marquis for time-travelling help, the *hmmm...* insights, and dimensional wisdom beyond my lived experience. I'm grateful to Tina Lee for the kindest of thoughts. And to Katel LeDû for helping me see all this through with compassion and poise.

This book wouldn't be what it became without the early feedback and pep talks of Marty Henderson and Geoffrey Pursell. I extend ceaseless thanks for your time and insight charting early tributaries of this content. To you Geoff, a lot of my open source journey is owed. You saw promise in me and then gave me a mirror.

Thanks to Bobby Stockstead for being a resolute sounding board and at times philosophical foil.

Thanks to Lina Jebara for asking the tough questions when no one else will.

To Brad Frost and Dave Olsen, thank you for supporting me with unflappable confidence, patience, and energy when I was just some punk forking shit because I didn't know PHP. That fork changed my life and made me a life-long open source believer.

Thank you to Alex Russell, Alice Goldfuss, Brad Frost, Duane O'Brien, Jenn Schiffer, Kat Marchán, Kelsey Hightower, and Shawn Wang for agreeing to attribute social media posts without direct links, in part due to instability of or exodus from the platform.

Unending gratitude to Andre "Staltz" Medeiros for relicensing the "Time Till Open Source Alternative" graphic especially for this book. Check out his Patreon @ https://patreon.com/andrestaltz

My humblest of thanks to Randall Munroe for granting permission to include his xkcd comic in the book. Your comics hold a lot of collective mindshare and I am glad for them to be present.

I try to pay forward several open source communities for their nurturing, collaborative atmosphere, but the Node.js community especially. Among many, Claudio Wunder, Antoine du Hamel, Jacob Smith, Wesley Todd, Rafael Gonzaga, Jean Burellier, and Robin Bender Ginn made me feel welcome from day one.

Monumental thanks to Abby Cabunoc Mayes for her time and insights into open source history and community-building.

Thanks to Luke Schlangen for his blessing to use the title "Approachable Open Source" for this book, which was not as novel as I'd thought.

Thanks to my father Jim Muenzenmeyer for detailing out the many ADA requirements building code must adhere to, and for your stories, disposition, and dreams. Long before this, you were my role model, in big ways and small. You gifted me my first ABA book when I was but a babe, how odd that I can now thank you within one (almost).

To Max, Jack, Ben, and Henry—you are the best of me and I cannot wait to see y'all grow up together in brotherhood. You put up with a lot of deferred time for me to write this. Show me this passage and we'll go do whatever you want.

The final word always goes to you, Megan. Maybe you'll see the back of the book too. Through all this and before and forward, it's because of and for you. You don't need to read between the lines with me, I love you.

1. https://www.dictionaryofobscuresorrows.com/
2. https://www.dictionaryofobscuresorrows.com/post/23536922667/sonder
3. https://www.linuxfoundation.org/blog/blog/a-summary-of-census-ii-open-source-software-application-libraries-the-world-depends-on
4. https://www.hbs.edu/faculty/Pages/item.aspx?num=65230
5. https://www.gutenberg.org/files/790/790-h/790-h.htm
6. https://www.computerhistory.org/revolution/personal-computers/17/312/2311
7. https://www.gnu.org/philosophy/free-sw.en.html
8. https://www.gnu.org/philosophy/free-sw.en.html#four-freedoms
9. https://opensource.org/history
10. https://opensource.org/definition/
11. https://norabateson.wordpress.com/about/
12. https://www.youtube.com/watch?v=3_VHsrz0kfc
13. https://medium.com/@jessitron/symmathecist-n-c728957ce71f
14. https://xkcd.com/2347/
15. https://github.com/facebook/jest/issues/13808
16. https://www.secondcity.com/how-to-say-yes-and/
17. https://github.com/facebook/create-react-app/issues/11756
18. https://www.redhat.com/en/blog/dont-lick-cookie
19. https://almanac.httparchive.org/en/2022/page-weight
20. https://github.com/moment/moment/issues/3979
21. https://github.com/zloirock/core-js/blob/master/docs/2023-02-14-so-whats-next.md
22. https://www.sciencedaily.com/releases/2018/04/180420122913.htm
23. https://peps.python.org/pep-0008/
24. https://patternlab.io/
25. https://bradfrost.com/blog/post/creative-exhaust/
26. https://bradfrost.com/blog/post/just/

27. https://maggieappleton.com/garden

28. https://unsplash.com/photos/yellow-flowers-on-the-garden-bY_q4VodUc0

29. https://beta.ada.gov/topics/intro-to-ada/

30. https://inclusive.microsoft.design/tools-and-activities/Inclusive101Guidebook.pdf

31. https://www.barrierfreegolf.com/

32. https://www.rampur.is/

33. https://www.phila.gov/2022-10-14-city-of-philadelphia-releases-statement-on-agreement-to-important-ada-class-action-settlement/

34. https://twitter.com/whitep4nth3r/status/1557092356138098699

35. https://twitter.com/flaviocopes/status/1417007331930423298

36. https://api.jquery.com/on/

37. https://developer.mozilla.org/en-US/docs/Learn/JavaScript/Building_blocks/Events#event_bubbling_and_capture

38. https://github.com/scottjehl/picturefill

39. https://developer.mozilla.org/en-US/docs/Web/HTML/Element/picture

40. https://www.apache.org/foundation/how-it-works.html#roles

41. https://github.com/nodejs/node/issues/42828

42. https://github.com/nodejs/node/blob/main/doc/contributing/issues.md#triaging-a-bug-report

43. https://github.com/11ty/eleventy/issues?q=label%3Aneeds-votes+sort%3Areactions-%2B1-desc+label%3Aenhancement

44. https://web.archive.org/web/20130202062627/http://archive.aneventapart.com/2012/atlanta/?/2012/atlanta/

45. https://www.lukew.com/ff/entry.asp?1491

46. https://docs.github.com/en/codespaces

47. https://www.allthingsopen.org/

48. https://events.linuxfoundation.org/

49. https://hacktoberfest.com/

50. https://hacktoberfest.com/participation/

51. https://anitab-org.github.io/open-source-day/open-source-day/

52. https://nodejs.org/en/blog/announcements/diving-into-the-nodejs-website-redesign

53. https://www.outreachy.org/

54. https://github.com/AllInOpenSource/All-In

55. https://ohshitgit.com/

56. https://flexboxfroggy.com/

57. https://cssgridgarden.com/

58. http://www.csszengarden.com/

59. https://stylestage.dev/

60. https://opencollective.com/

61. https://www.patreon.com/

62. https://github.blog/2022-02-02-new-sponsors-only-repositories-custom-amounts-and-more/

63. https://jestjs.io/blog/2020/05/05/jest-26#stay-safe

64. https://openssf.org/community/alpha-omega/

65. https://trends.builtwith.com/javascript/jQuery

66. https://commission.europa.eu/about-european-commission/departments-and-executive-agencies/informatics/eu-fossa-2-free-and-open-source-software-auditing_en

67. https://aworkinglibrary.com/writing/no-one-is-non-technical

68. https://allcontributors.org/

69. https://allcontributors.org/docs/en/emoji-key

70. https://npmjs.org/

71. https://bower.io/blog/2017/how-to-migrate-away-from-bower/

72. https://almanac.httparchive.org/en/2022/page-weight#page-weight-by-the-numbers

73. https://almanac.httparchive.org/en/2022/javascript

74. https://github.com/browserslist/browserslist

75. https://browsersl.ist/#q=%3E+0.5%25%2C+last+2+versions%2C+Firefox+ESR%2C+not+dead

76. https://developer.mozilla.org/en-US/docs/Web/CSS/@supports

77. https://momentjs.com/timezone/tests/

78. https://time.is/time_zone_news

79. https://en.wikipedia.org/wiki/Time_in_Indiana

80. https://developer.mozilla.org/en-US/docs/Web/JavaScript/Reference/Global_Objects/Intl/DateTimeFormat#specifications

81. https://cldr.unicode.org/

82. https://unicode-org.github.io/icu/userguide/datetime/timezone/

83. https://www.iana.org/time-zones

84. https://www.npmjs.com/package/size-limit

85. https://github.com/GoogleChrome/lighthouse-ci?

86. https://create-react-app.dev/

87. https://nextjs.org/

88. https://npm.anvaka.com/#/view/2d/react-scripts

89. https://npm.anvaka.com/#/view/2d/next

90. https://github.com/csstools/postcss-normalize/issues/58

91. https://javascript.plainenglish.io/is-promise-post-mortem-cab807f18dcc

92. https://github.com/nrwl/nx/issues/15986

93. https://blog.npmjs.org/post/141577284765/kik-left-pad-and-npm

94. https://snyk.io/blog/open-source-npm-packages-colors-faker/

95. https://snyk.io/blog/peacenotwar-malicious-npm-node-ipc-package-vulnerability/

96. https://www.mend.io/resources/blog/cybercriminals-targeted-users-of-packages-with-a-total-of-1-5-billion-weekly-downloads-on-npm/

97. https://snyk.io/blog/npm-dependency-confusion-attack-gxm-reference/

98. https://youmightnotneedjquery.com/

99. http://youmightnotneedjs.com/

100. https://www.reddit.com/r/pics/comments/yxu5e6/starbucks_is_over_an_hour_behind_on_their_first

101. https://codepen.io/bmuenzenmeyer/pen/zYaNNYZ/6947c6986803f47d9b2ee319ce2d59b7

102. https://bundlephobia.com/package/moment@2.29.4

103. https://momentjs.com/docs/#/-project-status/

104. https://bundlephobia.com/package/dayjs@1.11.5

105. https://codepen.io/bmuenzenmeyer/pen/XWYppKR/b9f7bc0568319a35df03bb51096f5054

106. https://sparkbox.com/foundry/maker_series_recap_dan_mall_2015

107. https://github.com/iamkun/dayjs/blob/dev/test/plugin/relativeTime.test.js

108. https://developer.mozilla.org/en-US/docs/Web/JavaScript/Reference/Global_Objects/Intl/RelativeTimeFormat

109. https://caniuse.com/?search=RelativeTimeFormat

110. https://codepen.io/bmuenzenmeyer/pen/RwJKKaq/e220ec08eb2300bd7e0888ae08ad33e8?editors=0010

111. https://8112310.fs1.hubspotusercontent-na1.net/hubfs/8112310/LF%20Research/Harvard%20Census%20II%20of%20Free%20and%20Open%20Source%20Software%20-%20Report.pdf

112. https://npmtrends.com/date-fns-vs-dayjs-vs-luxon-vs-moment

113. https://www.typescriptlang.org/

114. https://jsdoc.app/about-getting-started

115. https://css-irl.info/disentangling-frameworks/

116. https://developer.chrome.com/blog/smooshgate/

117. https://mootools.net/

118. https://tc39.es/proposal-flatMap/

119. https://262.ecma-international.org/10.0/#sec-array.prototype.flat

120. http://info.cern.ch/hypertext/WWW/TheProject.html

121. https://www.spacejam.com/1996/

122. https://github.com/tc39/proposal-flatMap/pull/56

123. https://github.com/facebook/react/pull/25516

124. https://www.npmjs.com/package/license-checker

125. https://choosealicense.com/licenses/gpl-3.0/

126. https://www.npmjs.com/package/express

127. https://semver.org/

128. https://docs.npmjs.com/cli/commands/npm-install

129. https://docs.npmjs.com/cli/configuring-npm/npmrc

130. https://blog.rafaelgss.dev/why-you-should-pin-actions-by-commit-hash

131. https://www.npmjs.com/package/axios

132. https://staltz.com/time-till-open-source-alternative.html

133. http://pdp-10.trailing-edge.com/decus_20tap3_198111/01/decus/20-0079/readme.txt.html

134. https://carbon.now.sh/

135. https://github.com/nvm-sh/nvm

136. https://github.com/actions/setup-node

137. https://docs.npmjs.com/cli/commands/npx

138. https://github.com/iamkun/dayjs#readme

139. https://en.wiktionary.org/wiki/pave_the_cowpath

140. https://readme.so/

141. https://documentation.divio.com/

142. https://github.com/get-alex/alex

143. https://github.com/badges/shields

144. https://vimeo.com/785089142/9c270e9a74

145. https://github.github.com/gfm

146. https://github.com/mermaid-js/mermaid

147. https://github.blog/2022-02-14-include-diagrams-markdown-files-mermaid/

148. https://www.npmjs.com/package/mermaid?activeTab=readme

149. https://github.com/community/community/discussions/16925

150. https://blocks.githubnext.com/

151. https://choosealicense.com/no-permission/

152. https://docs.github.com/en/site-policy/github-terms/github-terms-of-service#5-license-grant-to-ot her-users

153. https://www.gnu.org/licenses/copyleft.en.html

154. https://creativecommons.org/

155. https://creativecommons.org/faq/#can-i-apply-a-creative-commons-license-to-software

156. https://opensource.org/proliferation-report/

157. https://en.wikipedia.org/wiki/Contributor_License_Agreement

158. https://lerna.js.org/

159. https://github.com/lerna/lerna/pull/1616

160. https://www.splcenter.org/news/2022/03/23/family-separation-timeline

161. https://www.texasobserver.org/the-biden-administration-is-still-separating-kids-from-their-familie s/

162. https://github.com/lerna/lerna/pull/1619

163. https://github.com/lerna/lerna/issues/1620

164. https://github.com/lerna/lerna/issues/1622

165. https://github.com/lerna/lerna/issues/1634

166. https://github.com/lerna/lerna/pull/1626

167. https://github.com/lerna/lerna/issues/1630

168. https://github.com/lerna/lerna/pull/1633

169. https://github.com/lerna/lerna/issues/1635

170. https://github.com/lerna/lerna/issues/1636

171. https://www.hashicorp.com/blog/hashicorp-adopts-business-source-license

172. https://blog.sentry.io/introducing-the-functional-source-license-freedom-without-free-riding/

173. https://redis.com/blog/redis-adopts-dual-source-available-licensing/

174. https://en.wikipedia.org/wiki/Free-rider_problem

175. https://www.linkedin.com/feed/update/urn:li:activity:7194150798820347904

176. https://www.linuxfoundation.org/press/announcing-opentofu

177. https://www.linuxfoundation.org/press/linux-foundation-launches-open-source-valkey-communi ty

178. https://mikepennisi.com/blog/2020/you-may-finally-use-jshint-for-evil/

179. https://www.linuxfoundation.org/blog/open-source-and-the-cra-will-not-work

180. https://ethicalsource.dev/what-we-believe/

181. https://firstdonoharm.dev/build/

182. https://www.npr.org/2022/10/26/1131377513/museum-protests-famous-artworks-history

183. https://githubcopilotinvestigation.com/

184. https://sfconservancy.org/GiveUpGitHub/

185. https://opensource.guide/legal/

186. https://amandabrock.com/books/

187. https://en.wikipedia.org/wiki/Atomic_commit

188. https://www.conventionalcommits.org/en/v1.0.0/

189. https://docs.github.com/en/pull-requests/collaborating-with-pull-requests/incorporating-changes-from-a-pull-request/about-pull-request-merges

190. https://docs.npmjs.com/creating-and-publishing-unscoped-public-packages

191. https://www.contributor-covenant.org/

192. https://docs.github.com/en/communities/setting-up-your-project-for-healthy-contributions/adding-a-code-of-conduct-to-your-project

193. https://github.com/nodejs/admin/blob/main/Moderation-Policy.md

194. https://regents.umn.edu/sites/regents.umn.edu/files/2022-07/policy_student_conduct_code.pdf

195. https://sage.thesharps.us/2015/10/05/closing-a-door/

196. https://github.com/torvalds/linux/commit/8a104f8b5867c682d994ffa7a74093c54469c11f

197. https://ethicalsource.dev/blog/culture-war-open-source/

198. https://css-tricks.com/psa-linking-to-a-code-of-conduct-template-is-not-the-same-as-having-a-code-of-conduct/

199. https://abookapart.com/products/inclusive-design-communities

200. https://wordpress.org/openverse/image/92821c41-3621-4d24-a27f-a798c7454737

201. https://danmall.com/posts/its-easier-to-revise-than-create/

202. https://developer.mozilla.org/en-US/docs/Web/CSS/justify-content

203. https://css-tricks.com/snippets/css/a-guide-to-flexbox

204. https://codepen.io/

205. https://lodash.com/docs/

206. https://www.11ty.dev/docs/search/

207. https://www.algolia.com/for-open-source/

208. https://w3techs.com/technologies/details/cl-en-

209. https://www.hootsuite.com/pages/digital-trends-2021

210. https://www.wired.com/story/coding-is-for-everyoneas-long-as-you-speak-english/

211. https://www.cve.org/ResourcesSupport/AllResources/CNARules#section_8-2_cve_record_prose_description_requirements

212. https://blog.codinghorror.com/the-ugly-american-programmer/

213. https://www.theguardian.com/education/2023/jun/20/netherlands-seeks-curbs-on-english-language-university-courses

214. https://semver.org/

215. https://en.wikipedia.org/wiki/Bjarkam%C3%A1l

216. https://whc.unesco.org/en/list/737/

217. https://archiveprogram.github.com/arctic-vault/

218. https://www.seedvault.no/

219. https://github.com/bmuenzenmeyer/CompassRose

220. https://8112310.fs1.hubspotusercontent-na1.net/hubfs/8112310/LF%20Research/Harvard%20Census%20II%20of%20Free%20and%20Open%20Source%20Software%20-%20Report.pdf

221. https://hbr.org/2013/03/the-ideal-praise-to-criticism

222. https://www.reddit.com/r/Inkscape/

223. https://www.reddit.com/r/eleventy/

224. https://stackoverflow.com/tags

225. https://forum.vivaldi.net/

226. https://docs.github.com/en/discussions

227. https://docs.github.com/en/account-and-profile/managing-subscriptions-and-notifications-on-github/setting-up-notifications/configuring-notifications#dependabot-alerts-notification-options

228. https://docs.gitlab.com/ee/user/profile/notifications.html#x-gitlab-notificationreason

229. https://www.apache.org/foundation/how-it-works

230. https://github.com/theopensourceway/guidebook

231. https://www.theopensourceway.org/the_open_source_way-guidebook-2.0.html#_to_build_diverse_open_source_communities_make_them_inclusive_first

232. https://www.theopensourceway.org/the_open_source_way-guidebook-2.0.html#_step_one_stop_saying_youre_a_meritocracy

233. https://opensourcesurvey.org/2017

234. https://twitter.com/LeaVerou/status/1614245838468915200

235. https://www.linuxfoundation.org/research/the-2021-linux-foundation-report-on-diversity-equity-and-inclusion-in-open-source

236. https://modelviewculture.com/pieces/the-dehumanizing-myth-of-the-meritocracy

237. https://www.artima.com/weblogs/viewpost.jsp?thread=235725

238. https://capwatkins.com/blog/the-sliding-scale-of-giving-a-fuck

239. https://github.com/prettier/prettier/issues/7475#issuecomment-668544890

240. https://editorconfig.org/

241. https://eslint.org/

242. https://eslint.org/play/

243. https://prettier.io/

244. https://code.google.com/archive/p/js-test-driver/wikis/GettingStarted.wiki

245. https://github.com/pattern-lab/patternlab-node/actions/workflows/continuous-integration.yml

246. https://docs.github.com/en/pull-requests/collaborating-with-pull-requests/collaborating-on-repositories-with-code-quality-features/about-status-checks

247. https://github.com/sindresorhus/np

248. https://docs.npmjs.com/cli/v9/using-npm/scripts#life-cycle-scripts

249. https://typicode.github.io/husky/#/

250. https://github.com/avajs/ava/blob/main/docs/recipes/watch-mode.md

251. https://jestjs.io/docs/cli#--watch

252. https://nodejs.org/api/test.html#watch-mode

253. https://github.com/facebook/create-react-app/issues/11174

254. https://pages.cloudflare.com/

255. https://www.netlify.com/

256. https://vercel.com/

257. https://intuit.github.io/auto/

258. https://docs.npmjs.com/cli/configuring-npm/package-json#files

259. https://docs.npmjs.com/cli/commands/npm-deprecate

260. https://docs.npmjs.com/cli/commands/npm-dist-tag

261. https://www.hbs.edu/faculty/Pages/item.aspx?num=65230

262. https://www.reuters.com/article/us-autos-selfdriving-uber/self-driving-uber-car-kills-arizona-woman-crossing-street-idUSKBN1GV296

263. https://www.justice.gov/opa/pr/eight-individuals-charged-deceptive-trading-practices-executed-us-commodities-markets

264. https://slate.com/technology/2022/12/albania-cyberattack-iran-it-workers-arrested.html

265. http://lawreview.law.pitt.edu/ojs/lawreview/article/view/822

266. http://www.safetyresearch.net/Library/BarrSlides_FINAL_SCRUBBED.pdf

267. https://www.thelancet.com/journals/lancet/article/PIIS0140-6736(17)30957-1/fulltext

268. https://adactio.com/articles/18580

269. https://www.nasa.gov/johnson/HWHAP/the-overview-effect

270. https://www.hq.nasa.gov/office/pao/History/alsj/a410/AS8-14-2383HR.jpg

271. https://foldingathome.org/

272. https://www.bemyeyes.com/

273. https://www.openwheelchair.org/LicenseInfo

274. https://www.looseendsproject.org/

275. https://www.un.org/sustainabledevelopment/climate-action/

276. https://www.theverge.com/2021/3/2/22307671/glitch-workers-sign-historic-collective-bargaining-agreement-cwa

277. https://kickstarterunited.org/

278. https://nytimesguild.org/

279. https://abookapart.com/products/you-deserve-a-tech-union

280. https://www.jpl.nasa.gov/news/meet-the-open-source-software-powering-nasas-ingenuity-mars-helicopter

281. https://github.com/readme/featured/nasa-ingenuity-helicopter

282. https://tc39.es/

283. https://github.com/whatwg

284. https://github.com/w3c

285. https://drafts.csswg.org/css-nesting-1/

286. https://github.com/w3c/csswg-drafts/issues/7834

287. https://backstage.io/

288. https://innersourcecommons.org/

289. https://patterns.innersourcecommons.org/explore-patterns

290. https://en.wikipedia.org/wiki/Seven_Summits

291. https://bigmedium.com/ideas/boring-design-systems.html

292. https://www.softwaremaxims.com/blog/not-a-supplier

293. https://www.lukew.com/ff/entry.asp?1696

294. https://patterns.innersourcecommons.org/p/30-day-warranty

295. https://en.wikipedia.org/wiki/Not_invented_here

296. https://blog.alicegoldfuss.com/slapping-back-imposter-syndrome/

297. https://www.quora.com/Why-dont-paramedics-run-to-emergency-patients/answer/Ross-Cohen-1

298. https://daverupert.com/2021/12/sustaining-maintaining/

299. https://sustainoss.org/

300. https://sustainoss.org/podcasts/

301. https://chaoss.community/

302. https://www.youtube.com/watch?v=a8UET_HLvu0

303. https://www.pekovichwoodworks.com/shop/the-why-and-how-of-woodworking

304. https://brianmuenzenmeyer.com/i-maintainer

305. https://www.duaneobrien.com/following-up-on-my-github-universe-talk/

306. https://accelerator.github.com/

307. https://clerk.com/blog/zod-fellowship

308. https://twitter.com/kantrn/status/1511791378497384454

309. https://www.npmjs.com/package/minimist

310. https://arstechnica.com/security/2024/04/what-we-know-about-the-xz-utils-backdoor-that-almos t-infected-the-world/

311. https://humanewebmanifesto.com/

312. https://www.canr.msu.edu/news/wind_is_essential_to_natural_processes

About the Author

Photo credit Samantha Brinda Photography

Brian lives in Chanhassen, Minnesota but hails from Manitowoc, where it was always cooler by the lake, and is now a fantastic ice-breaker at parties. He has four boys that keep him bald. He does a little of everything. Sometimes that's called being a unicorn, duck, or a jack-of-all-trades, but he likes to land on the Seussical-form—Sneelock of the Circus!

With a career spanning many roles, from developer, UX team of one, product manager, analyst, and freelancer, Brian brings an experienced and broad approach to many disciplines. Him and his wife Megan's small business keeps them exhausted and grounded in customer delivery, innovation, and warrantee-voiding laser maintenance. He's been published in *Smashing Magazine*, *CSS-Tricks*, and led workshops at Web Design Day and the Node.js Collaborator Summit. Open source software threads into many aspects of his life, and has opened doors he'd never thought imaginable. It can do that for you too.

When not writing or working within open source software, Brian lives out programming tropes of drinking coffee and woodworking. He enjoys soccer, playing games with his sons, especially X-Wing or Chess, and never turns down a milkshake. He and Megan spend as much time outside as they can muster, often playing with their kids, deepening the pickleball rivalry on their makeshift court, chasing clouds, or digging up the yard.

Visit Brian at https://brianmuenzenmeyer.com/

www.ingramcontent.com/pod-product-compliance
Ingram Content Group UK Ltd.
Pitfield, Milton Keynes, MK11 3LW, UK
UKHW020627120926
15112UKWH00034B/83

9 798990 779808